ANNALS OF MATHEMATICS STUDIES
NUMBER 9

DEGREE OF APPROXIMATION BY POLYNOMIALS IN THE COMPLEX DOMAIN

BY

W. E. SEWELL

PRINCETON
PRINCETON UNIVERSITY PRESS
LONDON: HUMPHREY MILFORD
OXFORD UNIVERSITY PRESS
1942

Lithoprinted in U.S.A.
EDWARDS BROTHERS, INC.
ANN ARBOR, MICHIGAN
1942

PREFACE

The study of degree of approximation by polynomials in the complex domain is in its second phase. An extensive investigation has been made of the relation of (geometric) degree of convergence by polynomials, approximating functions analytic on a given closed set, to regions of analyticity on the one hand and to regions of uniform convergence on the other hand; the methods and results of this investigation are available in an admirable treatise by Walsh [1935]. The more delicate question of the relation between the behavior (continuity properties, asymptotic conditions, singularities, etc.) of the function on the boundary of the region of convergence, and degree of convergence on the given set of approximating polynomials, is not included in the above treatise. This new problem is the subject treated in the present work.

The analogous question for functions of a single real variable is classical, and has been studied especially by de la Vallée Poussin, Lebesgue, Bernstein, Jackson, and Montel. This involves approximation both by polynomials in the independent variable and by trigonometric sums. Since a polynomial in the complex variable is on the unit circle a trigonometric sum with the real variable chosen as arc length, classical results concerning trigonometric approximation can be applied at once to our present problems. Furthermore the theory of conformal mapping extends the results for the unit circle to a wide variety of regions.

The study of this problem in the complex domain is comparatively recent; the theorem that a function analytic in a Jordan region and continuous in the closed region can be uniformly approximated by polynomials was proved by Walsh in 1926. This theorem along with results and methods concerning regions of analyticity and degree of convergence is a necessary forerunner to a systematic treatment of the present problem.

The appearance of this book does not indicate that the study is completed, far from it, but certain parts of the problem have been solved and the general progress is such that a perspective is now obtainable. In spite of the comparatively few years devoted to an organized study of the question it is impossible to include all of the results in a single treatise. No attempt is made to present an encyclopedic account of the material; the scheme is rather to trace the progress in the study of the problem and to motivate and explain the material evolved. In fact the purpose of the book is not only to present theorems and their proofs but also to stimulate interest in the subject by pointing out the limitations as well as the extent of the existing methods and calling attention to numerous specific problems as yet unsolved.

Many new results are published here for the first time. Much of the material on approximation as measured by a line integral was developed during the preparation of this book; some of the results on Tchebycheff approximation have been extended and refined by the introduction of more powerful methods. Also many of the exercises, stated as theorems, have not previously appeared in the literature; suggestions are given as to their solution in many instances, especially where the methods developed in the text do not apply. At the end of each chapter these exercises are followed by a discussion of open problems.

As the title indicates, the treatment is restricted to approximation by polynomials. Interpolation is incidental and considered only in those cases where it follows naturally from the methods used in studying degree of approximation; many results on interpolating polynomials are included as exercises. Although emphasis is placed on regions bounded by analytic Jordan curves methods which can be readily extended are applied to more general situations. The generalized Lipschitz condition is used to describe the continuity of functions rather than modulus of continuity or generalized derivative. The entire treatment is intended to suggest that the theory studied is a living developing organism rather than an embalmed museum exhibit.

The results and methods of the present treatment can be applied in the study of the analogous theory of approximation to harmonic functions by harmonic polynomials In this connection some partial results have already been obtained by Walsh and the author [1940b].

References to the literature are inserted in the text. Figures in square brackets are dates and indicate particular works on which precise information is given in the Bibliography. In order to facilitate the use of the book, references are made to books rather than to original memoirs whenever possible, preferably to Walsh [1935]. The Bibliography makes no pretensions to completeness; further references to the literature are given for instance by Walsh [1935, 1935a] and by Shohat, Hille, and Walsh [1940].

The author has received invaluable aid in the preparation of this book. Professor J. L. Walsh has not only contributed to practically every page but has been a continual source of inspiration throughout the author's mathematical career. The new research presented in the present work was done largely at the author's own institution, the Georgia School of Technology, but the actual writing of the book was done at Harvard University in the Harvard College Library while the writer was on leave of absence from the Georgia School of Technology. The Julius Rosenwald Fund has been very generous both in grants freeing the author's time for the purpose of writing the book and in arranging for its publication. Mr. Henry Allen Moe has taken an interest in this project far beyond his capacity as a member of the Julius Rosenwald Fund and has encouraged the author throughout this undertaking. The typing of the manuscript and some of the proofreading were done by Miss Jeanne Le Caine. Professor J. H. Curtiss proofread the master copy for lithoprinting. To these sources the author is deeply grateful.

W. E. Sewell

CONTENTS

Page

Chapter I

PRELIMINARIES

§1.1. STATEMENT OF THE PROBLEMS; NOTATION. The domain under consideration throughout the entire treatise is the plane of the complex variable $z = x+iy$. We approximate only on a closed bounded set; we deal thus primarily with the plane of finite points although on occasion we need to study in an auxiliary capacity the plane extended by adjunction of a single point at infinity. The usual definitions of point set theory are assumed, but it is well to recall certain concepts. A Jordan arc is a one to one continuous transform of a line segment, that is, a point set represented by points (x,y) where $x = f(t)$, $y = g(t)$, $0 \leq t \leq 1$, the functions $f(t)$ and $g(t)$ being continuous and admitting a unique solution t for given (x,y). If the functions $f(t)$ and $g(t)$ are analytic and $|f'| + |g'| \neq 0$ the Jordan arc is said to be analytic. A Jordan curve is a one to one continuous transform of a circumference, that is a point set represented by points (x,y) where $x = f(\theta)$, $y = g(\theta)$, $0 \leq \theta \leq 2\pi$, $f(0) = f(2\pi)$, $g(0) = g(2\pi)$, the functions being continuous and admitting a unique solution θ for given (x,y). If the functions are analytic and $|f'| + |g'| \neq 0$ the Jordan curve is said to be analytic. A point set E is connected if any two points of E can be joined by a Jordan arc consisting entirely of points of E. A region is an open connected set. A Jordan region is a region of the finite plane bounded by a Jordan curve. The terms integrable and measurable are in the sense of Lebesgue unless otherwise noted.

A function of the form

$$(1.1.1) \qquad p_n(z) \equiv a_n z^n + a_{n-1} z^{n-1} + \cdots + a_0,$$

where a_n, a_{n-1}, ..., a_0 are arbitrary complex constants, is called a polynomial of degree n in z; throughout this book the degree of a polynomial is indicated consistently by its subscript. In (1.1.1) we do not assume $a_n \neq 0$, thus it is clear that a polynomial of degree m, $m < n$, is also a polynomial of degree n; a constant, for example, is a polynomial of every degree. We say that a sequence of polynomials $\{p_n(z)\}$, n = 0, 1, 2, ..., converges to a function f(z) defined on a set E in the sense of Tchebycheff, or approximates to f(z) on E in the sense of Tchebycheff, if

$$(1.1.2) \qquad |f(z) - p_n(z)| \leqq \epsilon_n, \ z \text{ on } E, \ \lim_{n\to\infty} \epsilon_n = 0.$$

If

$$\underset{z \text{ on } E}{\text{l.u.b.}} \ |f(z) - p_n(z)| = \epsilon_n,$$

the infinitesimal ϵ_n is a measure of the degree of convergence or the degree of approximation of $p_n(z)$ to f(z) on E in the sense of Tchebycheff, and the sequence ϵ_1, ϵ_2, ϵ_3, ... measures the degree of convergence of the sequence $\{p_n(z)\}$ to f(z) on E.

Let E, with boundary C be a closed limited point set in the z-plane and let the function f(z) be defined on E; we use the notation $\overline{C} = E$. Suppose the boundary C of E consists of one or several Jordan curves, and let f(z) be analytic merely in the interior points of E and continuous on E; here approximation on C and E are identical. Problem α denotes the study of the relation between the degree of convergence of certain sequences $p_n(z)$ to f(z) on E on the one hand and the continuity properties of f(z) on C on the other hand. Part I is devoted to a study of Problem α ; in particular we study Tchebycheff approximation in Chapter III. This problem has a direct analogue in the real domain, that is for E a segment of the axis of reals; these results,* which are now classical, are applied to the general case considered here. In the complex domain Walsh [1935 pp. 36-39] showed that $f(z)$ can be uniformly approximated on E (including the case of E a Jordan arc); recent contributions to Problem α are due primarily to Curtiss, Sewell, and Walsh and Sewell.

Now let E, with boundary C, be a closed limited set whose complement (with respect to the extended plane) K is connected and regular** in the sense that K possesses a Green's function $G(x,y)$ with pole at infinity; then the function

$$(1.1.3) \qquad w = \phi(z) = e^{G(x,y) + iH(x,y)}, \quad z = x+iy,$$

where $H(x,y)$ is conjugate to $G(x,y)$ in K, maps K conformally, but not necessarily uniformly, on the exterior of the unit circle, γ : $|w| = 1$, in the w-plane so that the points at infinity in the two planes correspond to each other. We denote by C_ρ the image, under the conformal map, in the z-plane of the circle $|\phi(z)| = |w| = \rho > 1$, that is to say C_ρ is the locus $G(x,y) = \log \rho > 0$ in K.

Let $f(z)$ be analytic on E; then there exists a greatest number ρ (finite or infinite) such that $f(z)$ is single valued and analytic at every point interior to C_ρ ; if E falls into several disconnected parts the function $f(z)$ defined on its various parts need not consist of the branches of a single monogenic analytic function. If $R < \rho$ and arbitrary there exist polynomials $p_n(z)$, $n = 0, 1, 2, \ldots$, such that

* See, e.g., Bernstein [1926], Jackson [1930], de la Vallée Poussin [1919].

**See, e.g., Walsh [1935, pp. 65, ff.].

(1.1.4) $$|f(z) - p_n(z)| \leqq \frac{M}{R^n}, \quad z \text{ on } E,$$

where M depends on R but not on n or z; when used in an inequality of the form (1.1.4) the letter M with or without subscripts shall hereafter represent a constant which may vary from inequality to inequality but is always independent of n and z. On the other hand there exist no polynomials $p_n(z)$ such that (1.1.4) is valid for z on E with $R > \rho$. A sequence $p_n(z)$, satisfying (1.1.4) for every $R < \rho$ and a suitable M is said to converge to f(z) on E <u>maximally</u>, or with the greatest geometric degree of convergence. Conversely if f(z) is defined on E and if for every n and every $R < \rho$ there exists $p_n(z)$ such that (1.1.4)is valid, then the function f(z) is analytic in the interior of C_ρ. This type of convergence has been fully investigated [Walsh, 1935] and the relation between degree of convergence and regions of analyticity thoroughly studied.

We consider in Part II a more delicate problem whose solution makes use of both the methods and results of the above study. Let f(z) be given analytic interior to some fixed C_ρ and let its continuity properties on or in the neighborhood of C_ρ be precisely described; we designate by <u>Problem β</u> the study of the relation between the degree of convergence of various sequences $p_n(z)$ to f(z) on E on the one hand and the continuity properties of f(z) on or in a neighborhood of C_ρ on the other hand; Chapter VI is devoted to a detailed study of this problem. More discrimination is required in investigating Problem β than was necessary in the study of the relation between geometric degree of convergence and regions of analyticity. The degree of convergence is not expressed in general in a formula as simple as $\epsilon_n = M/\rho^n$; in fact the right member of (1.1.4) is usually multiplied by a positive or negative power of n, $(M/\rho^n).n^\sigma$. Our treatment of the problem amounts to a study of the <u>relation</u> between the number σ and the behavior of f(z).

Special cases of this problem have been studied by Bernstein, Faber, and de la Vallée Poussin, among others; the bulk of the material treated in the present work is of recent origin and is due primarily to Walsh and the author.

For the sake of brevity and exposition we distinguish between two types of results in both Problem α and Problem β. A direct theorem is one in which the properties of $f(z)$ are in the hypothesis and the degree of convergence of $p_n(z)$ to $f(z)$ is the conclusion. An indirect theorem is in the converse direction, that is the degree of convergence of $p_n(z)$ to $f(z)$ is in the hypothesis and the properties of $f(z)$ form the conclusion.

Throughout this treatise we describe the properties of a continuous function $f(z)$ by a generalized Lipschitz condition or closely related inequality. Let α be a fixed number, $0 < \alpha \leq 1$; a function $f(z)$ defined on a set E satisfies on E a Lipschitz condition of this given order α provided

$$(1.1.5) \qquad |f(z_1) - f(z_2)| \leq L|z_1 - z_2|^{\alpha} ,$$

where z_1 and z_2 are arbitrary points of E, and L is a constant independent of z_1 and z_2. For $\alpha = 1$ we say simply that the function satisfies a Lipschitz condition, frequently omitting the qualifying phrase "of order unity." We denote by $f'(z)$, $f''(z)$, $f'''(z)$ the first, second, and third derivatives respectively of $f(z)$; $f^{(k)}(z)$ denotes the k-th, $k = 1, 2, \ldots, m$, derivative of $f(z)$, and $f^{(0)}(z)$ denotes $f(z)$. Let E be a closed limited set bounded by a Jordan curve C; we say that a function $f(z)$ belongs to the class $L(k, \alpha)$ on C if $f(z)$ is analytic in the interior points of E, is continuous on E, and $f^{(k)}(z)$ exists on C in the one-dimensional sense and satisfies a Lipschitz condition of order α on C; in $L(k, \alpha)$ the number α is fixed, $0 < \alpha \leq 1$, and k is an integer. We say that $f(z)$ belongs to the class

Log(k,1) on C if $f(z)$ is analytic in the interior points of E, is continuous on E, and $f^{(k)}(z)$ exists on C in the one-dimensional sense and satisfies the condition

$$(1.1.6) \qquad |f^{(k)}(z_1) - f^{(k)}(z_2)| \leq L|z_1 - z_2|\cdot|\log|z_1-z_2||,$$

where z_1 and z_2 are arbitrary points of C, and L is a constant independent of z_1 and z_2; in Log (k,1) the number k is an integer, and the condition is assumed to be satisfied merely for $|z_1 - z_2|$ sufficiently small. This latter restriction is obviously essential, for the second member of (1.1.6) reduces to zero if $|z_1 - z_2|$ is unity. We say that $f(z)$ belongs to the class L(k, α) on E provided the function $f(z)$ is analytic in the interior points of E, is continuous on E, and provided $f^{(k)}(z)$, defined on C in the one-dimensional sense and interior to C in the usual way, satisfies on E a Lipschitz condition of order α; we say that $f(z)$ belongs to the class Log (k,1) on E provided $f(z)$ is analytic in the interior points of E, is continuous on E, and provided $f^{(k)}(z)$ defined on C in the one-dimensional sense and interior to C in the usual way, satisfies condition (1.1.6) on E. We give in the next section a detailed discussion of the relation between one-dimensional derivatives on C and two-dimensional derivatives on $\overline{C}$, and also of the relation between inequalities (1.1.5) and (1.1.6) on C and those inequalities on E.

Let C be a Jordan arc; we say that a function $f(z)$ belongs to the class L(k, α) on C provided $f(z)$ is continuous on C and $f^{(k)}(z)$ exists on C in the one-dimensional sense and satisfies a Lipschitz condition of order α on C; we say that $f(z)$ belongs to the class Log (k,1) on C provided $f(z)$ is continuous on C and $f^{(k)}(z)$ exists on C in the one-dimensional sense and satisfies condition (1.1.6) on C.

The above classifications require no modification if C consists of a finite number of Jordan curves or arcs or both. We say, for example, that f(z) belongs to the class L(k, α) on C provided f(z) belongs to the class L(k, α) on each component of C.

In addition to Tchebycheff approximation we consider also approximation measured by a line integral. Let f(z) be defined and integrable on a set C consisting of a finite number of rectifiable Jordan curves and suppose that for each n, n = 0, 1, 2, ..., there exists $p_n(z)$ such that

$$\int_C \Delta(z)\, |f(z) - p_n(z)|^p |dz| = \epsilon_n, \quad \lim_{n\to\infty} \epsilon_n = 0,$$

where p is a fixed positive number and Δ(z) is a non-negative norm or weight function defined and integrable on C. Degree of convergence or approximation is defined in terms of ϵ_n precisely as in the case of Tchebycheff approximation. Such approximation is called approximation in the sense of least weighted p-th powers; the concept and terms also apply to surface integrals. Here the degree of approximation depends on p as well as on f(z) and C; it also depends on Δ(z) but in our study of the problem the generality of the weight function is restricted so as not to affect materially the degree of approximation. Of course we consider in this connection both Problem α and Problem β, the former in Chapter IV, and the latter in Chapter VII.

Throughout our treatment we simply write: there exists $P_n(z)$, rather than for each n, where n includes every positive integral value from 1 or 2 depending on the conditions, there exists a polynomial $P_n(z)$ of degree n in z; sometimes it is necessary in this connection to take n large in which case it is to be understood that n ranges through every positive integral value exclusive of those values less than a fixed large integer. It is necessary to exclude the values 0 and 1 of n if the function log n appears as a factor; if n to a positive power appears in the denominator it is meaningless to include the value 0;

if in the numerator the value 0 must be excluded unless the function is a constant. Of course we are interested in convergence, and these restrictions are not serious.

§1.2. LIPSCHITZ CONDITIONS AND DERIVATIVES. The classes $L(\kappa, \alpha)$ and Log $(k,1)$ defined in §1.1 are of fundamental importance in our present study; the present section is devoted to a detailed discussion of the properties of functions of these classes.

Before discussing the case of an arbitrary Jordan region we study the situation for the unit circle $\gamma: \ |z| = 1$, and establish some interesting inequalities with reference to Lipschitz conditions and derivatives. Let $f(z)$ be analytic in $|z| < 1$ and continuous on $|z| \leqq 1$; then on the circumference $f(e^{i\theta}) \equiv F(\theta)$ is a function of the real variable θ, where $z = re^{i\theta}$, $r \leqq 1$. It is easy to show that if $f(z)$ satisfies a Lipschitz condition of order α, $0 < \alpha \leqq 1$, in z on γ, then it satisfies a Lipschitz condition of that same order α in θ; conversely if $F(\theta)$ satisfies a Lipschitz condition of order α in θ then $f(z)$ satisfies the same condition in z on γ. These facts are important in the application of classical results of the real domain to approximation to functions of the complex variable.

In the theorems which follow we assume $f(z)$ analytic in $|z| < 1$ and continuous on $|z| \leqq 1$.

THEOREM 1.2.1. A necessary and sufficient condition that $f(z)$ belong to the class $L(0, \alpha)$ on γ is that

(1.2.1) $\quad |f'(re^{i\theta})| \leqq L\cdot(1-r)^{\alpha-1}, \ r < 1,$

where L is a constant independent of r.

Suppose $f(z)$ belongs to the class $L(0, \alpha)$ on γ; then by applying the principle of the maximum to the function

$$g(z) \equiv f(ze^{ih}) - f(z),$$

we have

(1.2.2) $$|f(re^{i(\theta+h)}) - f(re^{i\theta})| \leq Lh^{\alpha} ,$$

uniformly for $0 \leq r < 1$ and all θ; thus $f(re^{i\theta})$ satisfies a Lipschitz condition of order α in θ uniformly with respect to r. Hence for $\alpha = 1$ it follows from (1.2.2) that $f'(re^{i\theta})$ is uniformly bounded.

Now suppose $f(z)$ belongs to the class $L(0, \alpha)$, $0 < \alpha < 1$. Then we have

$$f'(re^{i\theta}) = \frac{1}{2\pi} \int_{-\pi}^{\pi} \frac{f(e^{i\phi})\, e^{i\phi} d\phi}{(e^{i\phi} - re^{i\theta})^2}$$

$$= \frac{e^{-i\theta}}{2\pi} \int_{-\pi}^{\pi} \frac{f(e^{i(\theta+\phi)})\, e^{i\phi} d\phi}{(e^{i\phi} - r)^2}$$

$$= \frac{e^{-i\theta}}{2\pi} \int_{-\pi}^{\pi} \frac{e^{i\phi}}{(e^{i\phi}-r)^2} [f(e^{i(\theta+\phi)}) - f(e^{i\theta})]\, d\phi.$$

Hence, if $1/2 \leq r < 1$

$$|f'(re^{i\theta})| \leq \frac{L_1}{2\pi} \int_{-\pi}^{\pi} \frac{|\phi|^{\alpha}}{|e^{i\phi} - r|^2}\, d\phi \leq L_2 \int_0^{\infty} \frac{\phi'^{\alpha}}{(1-r)^2+\phi'^2}\, d\phi'$$

$$\leq L_3 \, (1-r)^{\alpha - 1},$$

where L_3 is independent of r. Since $f'(z)$ is analytic for $|z| < 1$, its modulus will be dominated by $L_3(1-r)^{\alpha - 1}$ throughout the interior of the unit circle.

On the other hand suppose (1.2.1) is valid for $0 < \alpha \leq 1$. For $0 < h \leq 1$ we have

$$f(e^{i(\theta+h)}) - f(e^{i\theta}) = \int_{e^{i\theta}}^{e^{i(\theta+h)}} f'(z)dz$$

$$= \left[\int_{(1)} + \int_{(2)} + \int_{(3)} \right] f'(z)\, dz$$

$$= I_1 + I_2 + I_3$$

where the paths of integration are respectively: (1) the radius from $e^{i\theta}$ to $(1-h)e^{i\theta}$, (2) the circle $r = 1-h$ from $(1-h)e^{i\theta}$ to $(1-h)e^{i(\theta+h)}$, and (3) the radius from $(1-h)e^{i(\theta+h)}$ to $e^{i(\theta+h)}$. The integrals I_1 and I_3 are each less than

$$L \left[\int_{1-h}^{1} (1-r)^{\alpha-1} dr \right] \leqq L_1 h^\alpha .$$

Also it is clear that $I_2 \leqq L[1-(1-h)]^{\alpha-1} h = Lh^\alpha$, and the proof of the theorem is complete.

For Lipschitz conditions on $|z| = 1$ and on $|z| \leqq 1$ we have

THEOREM 1.2.2. If $f(z)$ satisfies a Lipschitz condition of order α in z (or in θ) on $|z| = 1$ then it belongs to the class $L(0, \alpha)$ on $|z| \leqq 1$.

The proof consists in establishing Lipschitz inequalities in θ and in r uniformly with respect to r and to θ respectively, and then applying the triangle inequality. From the hypothesis we have (1.2.2) just as above and hence a Lipschitz condition in θ uniformly with respect to r.

In the radial direction we let $\theta = 0$ without loss of generality; then we have by Theorem 1.2.1

$$|f(r+h) - f(r)| \leqq \int_r^{r+h} |f'(t)| dt \leqq L \int_r^{r+h} (1-t)^{\alpha-1} dt$$

If $r+h \leqq (1+r)/2$, then we have $h \leqq (1-r)/2$, and $(1-t) \geqq (1-r)/2$;
hence

$$\int_r^{r+h} (1-t)^{\alpha-1} dt < h(1-r)^{\alpha-1} < L_1 h\, h^{\alpha-1} = L_1 h^\alpha ;$$

on the other hand suppose $r+h \geqq (1+r)/2$, then we have $h \geqq (1+r)/2$, and

$$\int_r^{r+h} (1-t)^{\alpha-1}dt < L(1-r)^{\alpha} < Lh^{\alpha}.$$

Thus we have

$$(1.2.3) \qquad |f((r+h)e^{i\theta}) - f(re^{i\theta})| \leqq Lh^{\alpha}.$$

Now consider

$$|f(r_1r^{i\theta_1}) - f(r_2e^{i\theta_2})| \leqq |f(r_1e^{i\theta_1}) - f(r_2e^{i\theta_1})|$$
$$|f(r_2e^{i\theta_1}) - f(r_2e^{i\theta_2})|.$$

By (1.2.3) we have

$$|f(r_1e^{i\theta_1}) - f(r_2e^{i\theta_2})| \leqq L|r_1 - r_2|^{\alpha},$$

and by (1.2.2) we have

$$|f(r_2e^{i\theta_1}) - f(r_2e^{i\theta_2})| \leqq L'|\theta_1 - \theta_2|^{\alpha}.$$

Consequently

$$|f(r_1e^{i\theta_1}) - f(r_2e^{i\theta_2}) \leqq L(|r_1-r_2|^{\alpha} + |\theta_1-\theta_2|^{\alpha})$$
$$\leqq L_1(|r_1-r_2|+|\theta_1-\theta_2|)^{\alpha};$$

but for r_1 and r_2 uniformly bounded from zero, say not less than one half, we have

$$|r_1 - r_2| + |\theta_1 - \theta_2| \leqq L_1(|r_1-r_2| + |e^{i\theta_1} - e^{i\theta_2}|)$$
$$\leqq L_2(|r_1-r_2| + |r_1e^{i\theta_1} - r_1e^{i\theta_2}|)$$
$$\leqq L_3|r_1e^{i\theta_1} - r_2e^{i\theta_2}|,$$

by the triangle inequality. Thus we have

$$|f(r_1e^{i\theta_1}) - f(r_2e^{i\theta_2})| \leqq L_4|r_1e^{i\theta_1} - r_2e^{i\theta_2}|^{\alpha},$$

for $r_1 \geqq 1/2$, $r_2 \geqq 1/2$; by virtue of the analyticity of $f(z)$ in $|z| < 1$ the function $f(z)$ belongs to the class $L(0,1)$ in $|z| \leqq 1/2$. For $|z_1| > 1/2$ and $|z_2| < 1|2$ we have merely to introduce the point z_3 on the segment z_1z_2

and consider the two pairs of points z_1, z_3 and z_3, z_2.

$$\begin{aligned}|f(z_1) - f(z_2)| &\leqq |f(z_1) - f(z_3)| + |f(z_3) - f(z_2)| \\ &\leqq L_4|z_1 - z_3|^\alpha + L_5|z_3 - z_2| \\ &\leqq L_4|z_1 - z_3|^\alpha + L_5|z_3 - z_2|^\alpha \\ &\leqq L_6|z_1 - z_2|^\alpha .\end{aligned}$$

The proof of the theorem is complete.

In Theorem 1.2.2 we proved that if f(z) satisfies a Lipschitz condition of order α in θ for $|z| = 1$ then f(z) belongs to the class $L(0, \alpha)$ on $|z| \leqq 1$. We now prove that if f(z) satisfies a Lipschitz condition of order α in r then it belongs to the class $L(0, \alpha)$ on $|z| \leqq 1$.

THEOREM 1.2.3. If we have
(1.2.4) $|f(e^{i\theta}) - f(re^{i\theta})| \leqq L\,(1-r)^\alpha$, $0 < \alpha \leqq 1$, $0 \leqq r \leqq 1$,
where L is independent of r and θ, then f(z) belongs to the class $L(0, \alpha)$ on $|z| \leqq 1$.

Let $g(z) \equiv f(z)-f(\rho z)$ where $\rho > 0$ is fixed. By the principle of the maximum and (1.2.4) we see that $|g(z)| \leqq L(1-\rho)^\alpha$, $|z| \leqq 1$; consequently we have

$$(1.2.5)\quad |f(re^{i\theta}) - f(r\rho e^{i\theta})| \leqq L(1-\rho)^\alpha = \frac{L}{r^\alpha}(r-r\rho)^\alpha ,$$

$$0 < r \leqq 1.$$

Thus it follows from (1.2.5) that we have

$$(1.2.6)\quad |f(\rho_1 e^{i\theta}) - f(\rho_2 e^{i\theta})| \leqq L_1\,(\rho_1 - \rho_2)^\alpha ,$$

for ρ_1 and ρ_2 not less than some positive constant, say 1/2; furthermore, inequality (1.2.6) follows for ρ_1 and ρ_2 not greater than 1/2 by virtue of the analyticity of f(z) in $|z| \leqq 1/2$. Hence by the argument used under similar circumstances in the proof of Theorem 1.2.2 it follows that inequality (1.2.6) is valid for $0 \leq \rho_1 \leq 1$, $0 \leq \rho_2 \leq 1$, where L_1 is a constant independent of ρ_1, ρ_2, and θ. Inequality (1.2.6) shows that $f(re^{i\theta})$ satisfies a Lipschitz condition of order α in r uniformly with respect to θ. For $\alpha = 1$ it is clear from (1.2.6) that the derivative $f'(z)$ is uniformly bounded in $|z| < 1$; thus the conclusion of Theorem 1.2.3 is immediate for the case $\alpha = 1$.

For $\alpha < 1$ and $0 < h < 1$ we have

$$f(re^{i\theta}) - f((r-hr)e^{i\theta}) = \frac{1}{2\pi i} \int_0^{2\pi} [f(e^{i\phi}) - f((1-h)e^{i\phi})] \, P(r,\phi-\theta)\, d\phi,$$

where

$$P(r,\phi-\theta) = \frac{1 - r^2}{1 + r^2 - 2r\cos(\phi - \theta)}$$

is the Poisson kernel. We differentiate with respect to θ:

$$rf'(re^{i\theta}) - (r - hr)\, f'((r-hr)\, e^{i\theta})$$
$$= \frac{e^{-i\theta}}{2\pi i} \int_0^{2\pi} [f(e^{i\phi})-f((1-h)e^{i\phi})] \frac{\partial P}{\partial \theta}\, d\phi;$$

here

$$\left| \frac{\partial P}{\partial \theta} \right| = \left| \frac{2r(1-r^2)\sin(\phi-\theta)}{(1-2r\cos(\phi-\theta) + r^2)^2} \right| \leq \frac{A}{1-r},$$

where A is an absolute constant. Hence

$$(1.2.7) \quad |rf'(re^{i\theta}) - (r-hr)\, f'((r-hr)e^{i\theta})| \leq \frac{L_2 h^{\alpha}}{1-r}.$$

Let

$$r_n = 1 - 2^{-n}, \quad r_n(1 - h_n) = r_{n+1},$$

so that

$$h_n = \frac{2^{-n}}{1-2^{-n}} \leq L_3\, 2^{-n},$$

and let $r = r_n$, $h = h_n$, $n = 1,2,3, \ldots, N$, in (1.2.7).

Adding the resulting equations we have

$$|r_N f'(r_N e^{i\theta})| \leqq L_4 \; (2^{N(1-\alpha)} + 2^{(N-1)(1-\alpha)} + \ldots) + L_5$$

$$\leqq L_6 \; (2^{N(1-\alpha)}) \leqq L_7 \; (1-r_N)^{\alpha - 1}.$$

Since $1-r$ varies only by a factor 2 in the interval (r_{N-1}, r_N) we obtain by the principle of the maximum

$$|f'(z)| \leqq L_8(1-r)^{\alpha-1}, \; z = re^{i\theta},$$

where L_8 is independent of r and θ; thus the conclusion follows from Theorems 1.2.1 and 1.2.2.

In the above we have considered functions which satisfy Lipschitz conditions making use of the resulting asymptotic properties of the derivatives. It is interesting to study the relation between functions, their derivatives, and integrals in connection with the above radial asymptotic condition. Since the results are applicable to Problem β rather than to Problem α we postpone the general discussion; however the two following theorems are included here due to the close connection in method of proof and result with the preceding theorems. For derivatives we have

THEOREM 1.2.4. Let $f(z)$ be analytic in $|z| < 1$ and let

$$|f(z)| \leqq L(1-r)^{\beta}, \quad |z| = r < 1,$$

where L is independent of r. Then

$$|f'(z)| \leqq L_1(1-r)^{\beta-1}, \; |z| = r < 1,$$

where L_1 is independent of r.

For $\beta > 0$ the function vanishes and hence the theorem is trivial. Let $\beta \leqq 0$ and consider

$$f'(z_0) = \frac{1}{2\pi i}\int_{\delta} \frac{f(t)}{(t-z_0)^2}\,dt, \quad |z_0| < 1,$$

where δ is a circle in $|z| < 1$ with center at $z_0 = re^{i\theta}$ and radius $(1-r)/2$. Thus

$$|f'(z_0)| \leqq L(1-r)^{\beta} \int_{\theta}^{2\pi+\theta} \frac{d\phi}{|\rho e^{i\phi} - re^{i\theta}|^2}$$

$$\leqq L_2(1-r)^{\beta} \quad (\rho - r)^{-1} \leqq L_1(1-r)^{\beta-1},$$

and the proof is complete.

For integrals we have

THEOREM 1.2.5. Let $f(z)$ be analytic in $|z| < 1$ and let

$$|f'(z)| \leqq L(1-r)^{\beta-1}, \quad \beta < 0, \ |z| = r < 1,$$

where L is independent of r. Then

$$|f(z)| \leqq L_1(1-r)^{\beta}, \ |z| = r < 1,$$

where L_1 is independent of r.

Without loss of generality we take $\theta = 0$ and assume $f(0) = 0$; then we have

$$|f(r)| = \left|\int_0^r f'(t)dt\right| \leqq L\left| -\frac{(1-t)^{\beta}}{\beta}\Big]_0^r \right|$$

$$= L\left| -\frac{1}{\beta}\left[(1-r)^{\beta} - 1\right]\right|,$$

and since $\beta < 0$ we have

$$|f(r)| \leqq L_1(1-r)^{\beta},$$

as was to be proved. The case $\beta = 0$, which is excluded here, receives further attention in our study of Problem β (see §5.2).

By conformal mapping the above theorems extend to analytic Jordan curves. We have, for example

THEOREM 1.2.6. Let E, with boundary C, be a closed limited set whose complement is simply connected. Let f(z) belong to the class L(k, α), k ≥ 0, on C_ρ , ρ > 1. Then f(z) belongs to the class L(k, α) on $\overline{C}_\rho$ and

$$(1.2.8) \quad |f^{(k+1)}(z)| \leq L(\rho - r)^{\alpha - 1}, \quad 1 < r_0 \leq r < \rho, \quad z \text{ on } C_r,$$

where L is a constant independent of r.

Through the mapping function w = ϕ(z) (see §1.1) the function f(z) becomes F(w) which is analytic in 1 < |w| < ρ. This function can be separated into two functions $F_1(w)$, analytic in |w| > 1, and $F_2(w)$, analytic in |w| < ρ , by means of the Laurent development. Since $F_1(w)$ is analyti in |w| > 1 its derivatives are uniformly bounded in 1 $1 < r_0 \leq |w| \leq \rho$, where r_0 is arbitrary but fixed. The function $F_2(w)$ belongs to the class L(k, α) on |w| = ρ by virtue of the analyticity of C_ρ and the conditions on f(z). Hence by an obvious extension of Theorem 1.2.1 we have

$$|F_2^{(k+1)}(w)| \leq L_1(\rho - r)^{\alpha - 1}, \quad |w| = r < \rho , \quad r > r_0,$$

where L_1 is independent of r. Consequently $F^{(k+1)}(w)$ satisfies a similar inequality, thus due to the analyticity of the mapping function we have (1.2.8). Furthermore it is clear from the discussion above and Theorem 1.2.1 that $F^{(k)}(w)$ satisfies a Lipschitz condition of order α in $r_0 \leq |w| \leq \rho$, hence $f^{(k)}(z)$ satisfies a like condition in the closed ring bounded by C_ρ and C_r . Thus $f^{(k)}(z)$ satisfies a Lipschitz condition of order α on $\overline{C}_\rho$ due to its analyticity interior to C_ρ . The proof of the theorem is complete.

The statements and proofs of further theorems analogous to Theorems 1.2.1 to 1.2.5 for analytic Jordan curves are not difficult and are left to the reader. This completes our study of analytic Jordan curves and we are now ready to consider much more general boundaries. Our main result is a generalization of Theorem 1.2.2.

THEOREM 1.2.7. Let $f(z)$ be analytic interior to an arbitrary Jordan curve C and continuous on $\overline{C}$. For fixed α, $0 < \alpha \leqq 1$, let

$$\left| \frac{f(z) - f(z_0)}{(z - z_0)^{\alpha}} \right| \leqq L, \tag{1.2.9}$$

for all z and z_0, $z_0 \neq z$, on C, where L is independent of z_0 and z. Then (1.2.9) is valid for all z and z_0 on $\overline{C}$.

The proof of this theorem involves several steps. We first establish a result concerning a function analytic in $|w| < 1$ and satisfying certain inequalities on $|w| = 1$ except at the point $w = 1$; we establish an inequality on this function which leads by conformal mapping to (1.2.9) for z_0 fixed on C, z_0 corresponding to the point $w = 1$ under the transformation. With (1.2.9) established for z_0 fixed on C we complete the proof of Theorem 1.2.7 by allowing z and z_0 to vary over $\overline{C}$.

The basic lemma is

LEMMA 1.2.8 (a) Let $\Phi(w)$ be analytic in $|w| < 1$; also for each $\epsilon > 0$ and for each point on $|w| = 1$ different from $w = 1$ let a half neighborhood* exist such that for all w lying in this neighborhood we have $|\Phi(w)| \leqq 1 + \epsilon$.

(b) Let $\phi(w)$ be analytic in $|w| < 1$ and suppose $|\phi(w)| \geqq c > 0$, where c is a constant;

* A half neighborhood of w_0, $|w_0| = 1$, consists of the points ζ such that $|\zeta - w_0| < \delta > 0$, $|\zeta| \leqq 1$.

for each $\epsilon > 0$ let a sequence of points w_m exist which converges to $w = 1$ in an angle $|\operatorname{arc}(1-w)| \leq \eta < \pi/2$ independent of ϵ, and for which we have

$$(1.2.10) \qquad |\phi(w_m)| \leq e^{\frac{\epsilon}{|1-w_m|}}.$$

(c) Let $|\Phi(w)| \leq |\phi(w)|$ in a neighborhood of $w=1$, $|w| < 1$.

Then in $|w| < 1$ we have $|\Phi(w)| \leq 1$.

We base the proof of Lemma 1.2.8 on two additional lemmas.

LEMMA 1.2.9 Let $f(w)$ be analytic in $|w| < 1$; for each $\epsilon > 0$ and for each point on $|w| = 1$ different from $w = 1$, let $|f(w)| \leq 1 + \epsilon$ for all w in a neighborhood. Then a necessary and sufficient condition that $|f(w)|$ be bounded in $|w| < 1$ (in fact that $|f(w)| \leq 1$) is that for each $\epsilon > 0$ there exists a $\delta = \delta(\epsilon)$ such that*

$$(1.2.11) \qquad \liminf_{R \to 1} \int_{-\delta}^{\delta} \overset{+}{\log} |f(Re^{i\theta})| d\theta \leq \epsilon .$$

If in $|w| < 1$ we have $|f(w)| \leq M$, then obviously condition (1.2.11) is fulfilled.

On the other hand let $\epsilon > 0$ be given and let δ be the $\delta(\epsilon)$ in (1.2.11) belonging to ϵ. For $0 \leq r < R < 1$ it is well known that

$$\overset{+}{\log} |f(re^{i\phi})| \leq \frac{1}{2\pi} \int_{-\pi}^{\pi} \overset{+}{\log} |f(Re^{i\theta})| \cdot \frac{R^2 - r^2}{R^2 + r^2 - 2rR \cos(\theta - \phi)} d\theta.$$

* $\overset{+}{\log} |f| = \log |f|$, $|f| \geq 1$; $\overset{+}{\log} |f| = 0$, $|f| < 1$.

We divide the integral into sub-integrals over the intervals $-\pi$ to $-\delta$, $-\delta$ to δ , δ to π . By the Heine-Borel theorem there is an R_0 independent of δ , $0 < R_0 < 1$, such that for $R_0 \leqq R < 1$ and $-\pi \leqq \theta \leqq -\delta$ or $\delta \leqq \theta \leqq \pi$ we have

$$\overset{+}{\log} |f(Re^{i\theta})| \leqq \overset{+}{\log} (1+\epsilon) < \epsilon ;$$

therefore for fixed r and a suitable sequence* $R_\nu \uparrow 1 (\nu = 1, 2, \ldots)$ with $\nu \to \infty$ we have

$$\overset{+}{\log} |f(re^{i\phi})| \leqq \overset{+}{\log} (1+\epsilon) + \liminf_{R \uparrow 1} \frac{1}{2\pi} \int_{-\delta}^{\delta} \overset{+}{\log} |f(Re^{i\theta})| \cdot \frac{R+r}{R-r} d\theta \leqq \epsilon + \frac{\epsilon}{2\pi} \frac{1+r}{1-r} ,$$

whence the sufficiency follows by the arbitrariness of ϵ . The proof of Lemma 1.2.9 is complete.

LEMMA 1.2.10. Let u(w) be a non-negative potential function in $|w| < 1$. Let $w_0 = e^{i\alpha_0}$ be the vertex of an angle Ω bounded by two proper chords (not tangents) of $|w| = 1$ and let

(1.2.12) $\liminf u(w) \mid w - w_0 \mid = 0$

as w approaches w_0 in Ω . Then corresponding to each $\epsilon > 0$ there is a $\delta > 0$ such that

$$\limsup_{R \uparrow 1} \int_{\alpha_0 - \delta}^{\alpha_0 + \delta} u(Re^{i\theta})\, d\theta \leqq \epsilon .$$

Without loss of generality we take $w_0 = 1$. By hypothesis we have for a suitable sequence w_ν of Ω where w approaches 1,

$$\lim_{\nu \to \infty} (1-w_\nu)\, u(w_\nu) = 0.$$

*The notation $R_\nu \uparrow 1$ means $R_\nu \to 1$, $R_\nu \leqq R_{\nu+1} \leqq 1$.

For this sequence w_ν we have

$$(1.2.13)\quad |1-w_\nu| \leqq K(1-|w_\nu|),$$

where K is a constant depending only on Ω. Let $\epsilon > 0$ be given and choose a particular $w_\nu = \zeta$ of this sequence with the property that

$$u(\zeta)\,|1-\zeta| \leqq \frac{\epsilon}{\pi(K+1)^2}.$$

If we set $\delta = 1 - |\zeta|$, then for $|\zeta| = \rho < R < 1$ we have

$$|1-\zeta|u(\zeta) \geqq (1-\rho)\,u(\zeta) = \frac{1-\rho}{2\pi}\int_{-\pi}^{\pi} u(Re^{i\theta})\frac{R^2-\rho^2}{|\zeta - Re^{i\theta}|^2}d\theta$$

$$\geqq \frac{1-\rho}{2\pi}\int_{-\delta}^{\delta} u(Re^{i\theta})\;\frac{R^2-\rho^2}{|\zeta - Re^{i\theta}|^2}\,d\theta.$$

For $-\delta \leqq \theta \leqq \delta = 1-\rho$ it follows that

$$|\zeta - Re^{i\theta}| \leqq |\zeta - e^{i\theta}| \leqq |\zeta - 1| + |1-e^{i\theta}|$$

$$\leqq K(1-\rho) + |1-e^{i\delta}| \leqq (k+1)(1-\rho)$$

by virtue of (1.2.13), hence

$$\frac{\epsilon}{\pi(k+1)^2} \geqq \frac{(1-\rho)(R^2-\rho^2)}{2\pi(k+1)^2(1-\rho)^2}\int_{-\delta}^{\delta} u(Re^{i\theta})\,d\theta.$$

We have now merely to let R approach 1 to complete the proof of the lemma.

We proceed to prove Lemma 1.2.8. Since we may replace $\phi(w)$ by $\phi(w)/c$ we set $c = 1$ without loss of generality. By hypothesis for a suitably chosen $R < 1$ and $\delta > 0$ sufficiently small we have

$$\int_{-\delta}^{\delta} \overset{+}{\log}|\Phi(Re^{i\theta})|d\theta \leqq \int_{-\delta}^{\delta} \overset{+}{\log}|\phi(Re^{i\theta})|\,d\theta$$

$$(1.2.14)\qquad = \int_{-\delta}^{\delta} \log|\phi(Re^{i\theta})|\,d\theta.$$

Since $u(w) = \log |\phi(w)|$ is a non-negative potential function in $|w| <$ for which (1.2.12) is fulfilled, it follows from Lemma 1.2.10 that for each $\epsilon > 0$ there is a $\delta > 0$ such that

$$\lim_{R \uparrow 1} \sup \int_{-\delta}^{\delta} \log |\phi(Re^{i\theta})| d\theta \leq \epsilon .$$

Thus by (1.2.11) we see from Lemma 1.2.9 that $|\Phi(w)| \leq 1$ in $|w| < 1$. The proof of Lemma 1.2.8 is complete.

We turn now to the proof of Theorem 1.2.7. Let z_0 be fixed on C, and let $z = \Psi(w)$ map conformally $|w| < 1$ on the interior of C so that $\Psi(1) = z_0$. Since $\Psi(w) - \Psi(1) \neq 0$ in $|w| < 1$ each branch of $[\Psi(w) - \Psi(1)]^{-\alpha} = \chi(w)$ is analytic and single valued in $|w| < 1$. Let $F(w) = f(\Psi(w)) - f(\Psi(1))$ and form the functions

$$\Phi(w) = \frac{1}{L} F(w) \chi(w), \quad \phi(w) = \frac{M}{L} \chi(w),$$

where $|F(w)| \leq M$ for $|w| \leq 1$. Then $\Phi(w)$ is analytic in $|w| < 1$, continuous in $|w| \leq 1$, except possibly at $w = 1$, and on $|w| = 1$; $w \neq 1$, we see that $|\Phi(w)| \leq 1$; furthermore $|\Phi(w)| \leq |\phi(w)|$. Since $\Psi(w) - \Psi(1)$ is bounded and schlicht, the function $1/[\Psi(w) - \Psi(1)]$ is schlicht and bounded from zero; hence by the Verzerrungssatz* applied to the modulus of the mapping function we have

$$\left| \frac{1}{\Psi(w) - \Psi(1)} \right| \leq \frac{M_1}{(1 - |w|)^2} ,$$

where M_1 depends on $\Psi(w)$.

or

$$|\chi(w)| \leq \frac{M_2}{|1-w|^2}, \quad 0 \leq |w| < 1,$$

where M_2 depends on $\chi(w)$ and on the angle containing w. It follows that (1.2.10) is valid and hence Theorem 1.2.7 holds for z_0 fixed on C.

For z and z_0 both on $\bar{C}$ and $\alpha = 1$ we form the function

$$F(z, z_0) = \frac{f(z) - f(z_0)}{z - z_0} , \quad z \neq z_0$$

* See, e.g., Pólya and Szegö [1925], II, pp. 27, 202.

$$F(z_0, z_0) = f'(z_0),$$

provided $f'(z_0)$ exists. For z_0 fixed in the interior of C this function is bounded in a neighborhood of z_0, is analytic for z interior to C even for $z = z_0$, and is continuous for z on $\bar{C}$. But for z_0 on C and z interior to C we know from the above discussion that

$$(1.2.15) \qquad |F(z, z_0)| \leq L,$$

hence $|F(z, z_0)| \leq L$ for z_0 interior to C and z on $\bar{C}$. Thus inequality (1.2.9) is valid for z and z_0 both on C, one of the points on C and the other interior to C, and for both points interior to C; hence we have completed the proof for $\alpha = 1$.

For $\alpha < 1$ we consider the function

$$P(z, z_0) = \log \left| \frac{f(z) - f(z_0)}{(z - z_0)^\alpha} \right| .$$

For z_0 fixed interior to C this function is single valued and harmonic interior to C except for $z = z_0$, where we naturally have

$$\lim_{z \to z_0} \left| \frac{f(z) - f(z_0)}{(z - z_0)^\alpha} \right| = 0,$$

and except at those points z for which $f(z) = f(z_0)$, but this set of points has no limit point interior to C (we exclude the trivial case of $f(z) \equiv f(z_0)$). At a point of this set the function $P(z, z_0)$ becomes negatively infinite; also if z interior to C approaches z_1 on C for which $f(z_1) = f(z_0)$ the function $P(z, z_0)$ becomes negatively infinite. But if z interior to C approaches any other point z_1 on C the function $P(z, z_0)$ remains continuous in the two-dimensional sense and hence its value is not greater than log L. Since the superior limit of $P(z, z_0)$ as z interior to C approaches a point of C is not greater than log L, by the principle of the maximum for harmonic functions we have

$$P(z, z_0) \leqq \log L,$$

for z interior to C, z_0 fixed interior to C, wherever $P(z, z_0)$ is defined. Thus inequality (1.2.9) is valid for z and z_0 both on C, for one of the points interior to C and the other on C, and for both points interior to C. The proof of Theorem 1.2.7 is complete.

Thus we see that if a function f(z) belongs to the class $L(0,\alpha)$ on C, a Jordan curve, then f(z) belongs to the class $L(0,\alpha)$ on $\overline{C}$. We establish a similar result for functions belonging to the class Log (0,1).

THEOREM 1.2.11. Let f(z) be analytic interior to an arbitrary Jordan curve C and continuous on $\overline{C}$; also let

$$(1.2.16) \quad |f(z) - f(z_0)| \leqq L|z - z_0||\log|z-z_0||,$$

for z and z_0 on C, $z \neq z_0$, with $|z-z_0|$ sufficiently small, where L is a constant independent of z and z_0. Then (1.2.16) is valid for all z and z_0 on $\overline{C}$ for $|z-z_0|$ sufficiently small.

As mentioned in §1.1 the condition that $|z-z_0|$ be sufficiently small is inserted to rule out the possibility of its being unity thus causing the right hand side of (1.2.16) to vanish; we might require that $|z-z_0|$ be not greater than 1/2. Of course if inequality (1.2.16) holds for $|z-z_0|$ sufficiently small then it holds for all z and z_0 on $\overline{C}$ other than in neighborhoods of points z_1 such that $|z_1 - z_0| = 1$. In the proof we assume without real loss of generality that C is contained in a circle of radius 1/4; the method of proof is similar to that of Theorem 1.2.7.

Let z_0 be fixed on C and form the function

$$F(z, z_0) = \frac{f(z) - f(z_0)}{(z-z_0) \log (z-z_0)},$$

where a particular branch of $\log (z-z_0)$ is chosen. Since the branch point of the logarithmic function is on the boundary, the function $F(z, z_0)$ is analytic interior to C, continuous on $\bar{C}$ except possibly at the point z_0, and $|F(z, z_0)| \leqq L$, z and z_0 on C, $z \neq z_0$, since $|\log(z-z_0)| \geqq |\log |z-z_0||$. In Lemma 1.2.8 we set

$$\Phi (w) = \frac{1}{L} \frac{f(\Psi (w))- f(\Psi (1))}{[\Psi (w)- \Psi (1)] \log [\Psi (w)- \Psi (1)]}$$

and

$$\phi (w) = \frac{M}{L} \frac{1}{\psi (w) - \psi (1)},$$

where M is an upper bound of

$$\frac{f(\psi (w)) - f(\psi (1))}{\log [\psi (w) - \psi (1)]}$$

in $|w| < 1$; such a number M exists because the function approaches zero as z approaches z_0 and is continuous elsewhere. Since the hypothesis of the lemma is satisfied we see by the conclusion that (1.2.16) is valid for z_0 fixed on C and z on $\bar{C}$.

Now let z_0 be a point interior to C. The function $F(z, z_0)$ is no longer analytic interior to C because the branch point is on C and the function therefore is multiple valued. However on the universal covering surface of C with z_0 deleted this function is analytic and approaches zero as z approaches z_0 along any path interior to C; thus it cannot approach its least upper bound in a neighborhood of z_0. Consequently $|F(z, z_0)| \leqq L$, z on $\bar{C}$, z_0 interior to C, by the principle of the maximum. The conclusion of the theorem now follows as above for all z and z_0 on $\bar{C}$.

Thus for the function itself we see that inequalities

on C imply certain inequalities on $\overline{C}$. We turn now to the derivatives of the function.

Let C be a Jordan curve; let f(z) be analytic in C and continuous on $\overline{C}$. At a point z_0 in the interior of C the derivative $f'(z_0)$ is uniquely defined due to the analyticity of f(z) in a two-dimensional neighborhood of z_0. If z_0 is a point on C there are <u>three</u> different limits which may be appropriately considered in studying $f'(z_0)$. The one-dimensional derivative of f(z) on C is

$$(1.2.17) \qquad \lim_{z\to z_0,\ z \text{ on } C} \frac{f(z) - f(z_0)}{z - z_0},$$

provided this limit exists; this definition is valid for a function defined merely on the set C which may be an arc as well as a curve. The two-dimensional derivative is

$$(1.2.18) \qquad \lim_{z\to z_0,\ z \text{ on } \overline{C}} \frac{f(z) - f(z_0)}{z - z_0},$$

provided this limit exists. Finally

$$(1.2.19) \qquad \lim_{z\to z_0,\ z \text{ on } \overline{C}} f'(z)$$

may be considered provided, of course, $f'(z)$ and $f'(z_0)$ exist; here $f'(z)$ may be defined either by (1.2.17) or by (1.2.18). Theorems on approximation require as hypotheses existence and continuity of such limits as (1.2.17), (1.2.18), (1.2.19), so it is important for us to consider the general question of equivalence among these three limits. Of course similar questions arise for higher derivatives but extension is immediate by iteration.

THEOREM 1.2.12. Let f(z) be analytic interior to an arbitrary Jordan curve C and continuous on $\overline{C}$; also let f(z) have a derivative $f'(z)$ in the one-dimensional sense at each point z_0 of C and let

$$(1.2,20)\qquad \left|\frac{f(z) - f(z_0)}{z - z_0}\right|$$

be uniformly bounded for all z and z_0 on C. Then for every z_0 on C we have

$$(1.2.21)\qquad \lim_{z \to z_0,\, z \text{ on } \bar{C}} \frac{f(z) - f(z_0)}{z - z_0} =$$

$$\lim_{z \to z_0,\, z \text{ on } C} \frac{f(z) - f(z_0)}{z - z_0} = f'(z_0);$$

that is to say the two-dimensional derivative $f'(z_0)$ exists on C and is equal to the one-dimensional derivative $f'(z_0)$.

By Theorem 1.2.7 we know that (1.2.20) is uniformly bounded for z_0 fixed on C and z on $\bar{C}$. Since the function

$$\frac{f(z) - f(z_0)}{z - z_0}$$

is uniformly bounded and approaches one and the same limit when z on C approaches z_0 from either direction, it follows from a theorem of Lindelöf [1915] that this limit exists in the two-dimensional sense.

As a part of the hypothesis of Theorem 1.2.12 we assumed the boundedness of the difference quotient; the next theorem shows that under slight restrictions on C this boundedness is implied by the derivative conditions.

THEOREM 1.2.13. Let C be a rectifiable Jordan curve with the property that there exists a number $A \geq 1$ such that if r denotes the distance $|z_1 - z_2|$ between any two points of C and s denotes the arc-length of the shorter (or either if of same length) arc of C joining z_1 and z_2, we have

(1.2.22) $\frac{s}{r} \leqq A.$

Let $f(z)$ be continuous on C and have a bounded derivative on C in the one dimensional sense; $|f'(z)| \leqq M$. Then for z_1 and z_2 on C we have

$$\left| \frac{f(z_1) - f(z_2)}{z_1 - z_2} \right| \leqq MA.$$

Let $z = \psi(w)$ map the circle $|w| = 1$ on C. Then

$$\int_{z_1}^{z_2} f'(z)dz = \int_{w_1}^{w_2} f'(\psi(w)) \cdot \psi'(w)dw,$$

where the path of the first integral is along C and the second along the corresponding arc of the circle $|w| = 1$ or with $w = e^{i\theta}$ along the axis of the real variable θ. By (1.2.22) the derivative $\psi'(w)$ is uniformly bounded and by hypothesis $f'(z)$ is uniformly bounded, hence we have*

$$f(z_2) - f(z_1) = \int_{z_1}^{z_2} f'(z)dz.$$

Thus

$$|f(z_1) - f(z_2)| \leqq M \int_{z_1}^{z_2} |dz| \leqq M|\text{ arc } \widehat{z_1 z_2}| \leqq MA|z_1 - z_2|.$$

We now prove a theorem on the continuity of the derivative in the two-dimensional sense.

THEOREM 1.2.14. Let $f(z)$ be analytic interior to an arbitrary Jordan curve C and continuous on $\overline{C}$; also let $f(z)$ have a continuous derivative $f'(z)$ in the one-dimensional sense on C and let (1.2.20) be uniformly bounded for z and z_0 on C. Then the one-dimensional derivative $f'(z)$ is also the two-dimensional derivative; furthermore for z_0 on C we have

* See Saks [1937], p. 207.

$$\lim_{z \to \zeta_0,\, z \text{ on } \overline{C}} f'(z) = f'(z_0);$$

that is to say the function $f'(z)$ is continuous in the two-dimensional sense on $\overline{C}$.

Let $z = \psi(w)$ map $|w| < 1$ conformally on the interior of C. Let $\delta > 0$ be fixed and form the function

$$(1.2.23) \qquad F(w.\delta) = \frac{f[\psi(e^{i\delta} w)] - f[\psi(w)]}{(e^{i\delta} w) - \psi(w)},$$

which is analytic in $|w| < 1$ and continuous on $|w| \leq 1$; thus $F(w, \delta)$ can be represented in $|w| < 1$ by the Poisson integral

$$(1.2.24 \quad F(w, \delta) = \frac{1}{2\pi} \int_{-\pi}^{\pi} F(e^{i\phi}. \delta) \frac{1 - r^2}{1 + r^2 - 2r \cos(\phi - \delta)} d\phi,$$

where $w = re^{i\theta}$. By definition (1.2.23) it is clear that for w fixed in $|w| < 1$

$$\lim_{\delta \to 0} F(w, \delta) = f'(z)]_{z = \psi(w)};$$

this equation is also valid for w fixed on $|w| = 1$ by hypothesis, where $f'(z)$ represents the one-dimensional derivative taken along C:

$$\lim_{\delta \to 0} F(e^{i\theta} \delta) = f'[\psi(e^{i\theta})].$$

The function $F(e^{i\theta}, \delta)$ is continuous on $|w| = 1$ by hypothesis . Since $F(e^{i\theta}, \delta)$ is uniformly bounded in modulus for all δ and all θ it follows from the Osgood-Lebesgue theorem that the limit of the integral is the integral of the limit as δ approaches zero; hence

$$f'[\psi(w)] = \frac{1}{2\pi} \int_{-\pi}^{\pi} f'[\psi(e^{i\theta})] \frac{1 - r^2}{1 + r^2 - 2r \cos(\phi - \theta)} d\phi.$$

This is the representation of the function $f'[\psi(w)]$ by

the Poisson integral, and since $f'[\psi(e^{i\theta})]$ is a continuous function of θ these functional values are taken on continuously in the two-dimensional sense by $f'[\psi(w)]$:

$$\lim_{w\to e^{i\theta},\ |w|<1} f'[\psi(w)] = f'[\psi(e^{i\theta})].$$

Hence

$$\lim_{z\to z_0,\ z \text{ interior to } C} f'(z) = f'(z_0),$$

and also for z on C. The proof of the theorem is complete

The corresponding results for higher derivatives follow immediately by induction.

THEOREM 1.2.15. Let $f(z)$ be analytic interior to an arbitrary Jordan curve C and continuous on $\overline{C}$; also let $f^{(k)}(z)$ exist in the one-dimensional sense and be continuous on C, and let

$$(1.2.25)\qquad \left|\frac{f^{(j)}(z) - f^{(j)}(z_0)}{z - z_0}\right|, \quad j = 0, 1, 2, \ldots, k - 1,$$

z and z_0 on C, be uniformly bounded. Then the two-dimensional derivatives $f'(z)$, $f''(z)$, ..., $f^{(k)}(z)$ exist on C and are continuous in the two-dimensional sense on $\overline{C}$. Moreover if $f(z)$ belongs to the class $L(k,\alpha)$ or to the class Log $(k,1)$ on C, then $f(z)$ belongs to the class $L(k,\alpha)$ or to the class Log $(k,1)$ on $\overline{C}$.

It is well to recall that by Theorem 1.2.13 the boundedness of (1.2.25) is implied by the other conditions on $f(z)$ provided the chord and arc of C are infinitesimals of the same order*; this, of course, is not true for curves with cusps.

Theorems 1.2.1 to 1.2.5 are due to Hardy and Littlewood [1932]. Theorem 1.2.6 is due to Walsh and Sewell [1941]. Warschawski [1934] proved Lemmas 1.2.8, 1.2.9, 1.2.10, and Theorem 1.2.7 for z_0 fixed on C; the present form of Theorem 1.2.7 as well as Theorems 1.2.12 to 1.2.15 is given by Walsh and Sewell [1940].

§1.3. EXERCISES. 1.3.1 Let f(z) be analytic in a Jordan region R and satisfy a Lipschitz condition of order greater than unity in R. Then f(z) is a constant.

1.3.2. If f(z) satisfies a Lipschitz condition of order $\alpha > 0$ on a set E then it satisfies there a Lipschitz condition of every order $0 < \beta < \alpha$ on E.

1.3.3. Let f(z) satisfy a Lipschitz condition of order α on an analytic Jordan curve C; let f(a) = 0, where a is a point of C. Then for $\beta < \alpha$ the integral

$$\int_a^z (z - x)^{-\beta} f(x)\, dx$$

is bounded in modulus, where the path of integration is along the shorter arc of C joining a to z. [Sewell, 1937].

1.3.4. Let f(z) be defined on an analytic Jordan curve C; let f(a) = 0, where a is a point of C. Also let

$$\frac{d}{dz} \left[\int_a^z (z-x)^{-\alpha} f(x) dx \right], \quad 0 < \alpha \leqq 1,$$

be bounded. Then f(z) satisfies a Lipschitz condition of order α on C. [Sewell, 1937].

---

* Two infinitesimals x and y are said to be of the same order provided |x/y| is bounded above and below by two positive constants; x and y are said to be equivalent infinitesimals provided lim x/y = 1.

1.3.5. Let $f(z)$ belong to the class $L(k, \alpha)$, $k \geq 1$, on an analytic Jordan curve C. Then $f'(z)$ belongs to the class $L(k-1, \alpha)$ on C. Give sufficient conditions on the Jordan curve C, other than analyticity, for the truth of this proposition.

1.3.6. Under the hypothesis of Ex. 1.3.5 let

$$F(z) = \int_{z_0}^{z} f(z)dz,$$

where the point z_0 is fixed on $\overline{C}$ and the path of integration is along permissible arcs in $\overline{C}$. Then $F(z)$ belongs to the class $L(k+1, \alpha)$ on C.

1.3.7. Prove Theorem 1.2.7 for C an analytic Jordan curve by means of Theorem 1.2.2 and conformal mapping.

§1.4. DISCUSSION. We see by Theorem 1.2.7 that if a function analytic in a Jordan region and continuous in the corresponding closed region satisfies a Lipschitz condition of order α on the boundary then it satisfies that condition in the closed region; the possibility of extending this result to more general regions is worthy of investigation. In Theorem 1.2.13 we have a sufficient condition on C that a continuous derivative imply a bounded difference quotient; there arises the obvious question of a necessary condition.

Throughout §1.2 we confined our attention to Lipschitz conditions and pseudo-Lipschitz conditions, so to speak; more general continuity conditions remain to be studied. In a sense Theorem 1.2.11 is a modification of Theorem 1.2.7; further modifications are certainly worthy of investigation. A natural first step is to raise the logarithm to a power; also functions other than logarithmic might well be considered.

Another important continuity condition is the

Riemann-Liouville generalized derivative*. It has been shown [Sewell, 1937] that for a Jordan curve satisfying mild conditions a Lipschitz condition of order α implies a bounded generalized derivative of every order $\beta < \alpha$; in the converse direction, however, a relation has been established only for curves analytic or piecewise analytic. A successful study of this latter problem would lead immediately to some new results on approximation as well as being of interest per se.

We mention also an even more general condition used in describing the continuity of a function. Let $f(z)$ be defined on a set E. Let the function $\omega(\delta)$ be defined for $\delta \geq 0$ and let $\lim_{\delta \to 0} \omega(\delta) = 0$; we say that the modulus of continuity of $f(z)$ on E is $\omega(\delta)$ provided the inequality $|z_1 - z_2| \leq \delta$, where z_1 and z_2 are arbitrary points of E, implies the inequality $|f(z_1) - f(z_2)| \leq \omega(\delta)$. Suppose C is a Jordan curve and $f(z)$ is analytic in the interior of C and continuous on $\overline{C}$. What is the most general function $\omega(\delta)$ for which we have the following conclusion: if $f(z)$ has the modulus of continuity $\omega(\delta)$ on C then $f(z)$ has this same (admitting a multiplicative constant) modulus of continuity $\omega(\delta)$ on $\overline{C}$? For what moduli of continuity $\omega(\delta)$ is this true if C is an analytic Jordan curve? More generally what is the relation between the properties of the modulus of continuity $\omega(\delta)$ and the continuity properties of the curve C for which this conclusion holds?

*See Riemann [1892]; Liouville [1832]; Hardy and Littlewood [1928].

PART I. PROBLEM α

Chapter II

POLYNOMIAL INEQUALITIES

§2.1. A POLYNOMIAL AND ITS DERIVATIVES. In the study of the continuity properties of a function it is clear that the existence and properties of the derivative play an important role. Suppose f(z) is defined on a set E and suppose the sequence $p_n(z)$ converges to f(z) on E; then the study of the sequence of derivatives $p_n'(z)$ may enable us to establish the existence and even certain properties of the derivative f'(z) of f(z). In the present section we investigate the bound on a set E of the derivative $P_n'(z)$ of a polynomial $P_n(z)$ implied by a given bound on the polynomial itself.

Our method requires a study of the distance from C to C_ρ (in the notation of §1.1). It is sufficient for the applications to assume the set E, with boundary C, closed, and limited, with a complement K which is simply connected; the case of a set E which is a single point is excluded. Now let P be a point of C and denote by $d(P,C_\rho)$ the minimum distance from P to C_ρ; then we define the distance $d(C,C_\rho)$ as the minimum of $d(P,C_\rho)$ as P ranges over C. It is obvious that $d(C.C_\rho)$ is a function of ρ which depends on the properties of the boundary C.

If C is an analytic Jordan curve we know that the quantity

$$(2.1.1) \qquad \left| \frac{\psi(w_1) - \psi(w_2)}{w_1 - w_2} \right| ,$$

$|w_1| \geqq 1$, $|w_2| \geqq 1$, is uniformly bounded from zero; thus $d(C, C_\rho) \geqq K(C)(\rho - 1)$, since $|w_1 - w_2| \geqq (\rho - 1)$ if $|w_1| = \rho$ and $|w_2| = 1$. In fact if C is such that (2.1.1) is uniformly bounded from zero for $|w_1| = \rho$, $|w_2| = 1$ we have $d(C, C_\rho) \geqq K(C)(\rho - 1)$.

Let C be a Jordan curve consisting of a finite number of analytic Jordan arcs meeting in corners of exterior angular openings $\mu_\nu \pi$, $0 < \mu_\nu \leqq 2$, $\nu = 1, 2, \ldots, \lambda$. If P is a point, not a corner, we have by virtue of the analyticity of the curve C at P the inequality $d(P, C_\rho) \geqq K(P)(\rho - 1)$, where K(P) depends on the point P. Now let $z_0 = \psi(w_0)$ be a corner of exterior opening $\mu\pi$, then Osgood and Taylor [1913] have shown* that, in a neighborhood of w_0,

$$(2.1.2) \qquad \psi(w) - \psi(w_0) = (w - w_0)^\mu \cdot g(w),$$

where g(w) is analytic in a neighborhood of w exterior to the unit circle, and is continuous in the corresponding closed neighborhood. Thus $d(z_0, C_\rho) \geqq K(z_0)(\rho - 1)^\mu$. Let $t = 1$ if $\mu \leqq 1$ and $t = \mu$ if $\mu > 1$, and consider the quantity

$$(2.1.3) \qquad \left| \frac{\psi(w_1) - \psi(w_2)}{(w_1 - w_2)^t} \right|, \quad |w_1| \geqq 1, \ |w_2| = 1,$$

for w_1 and w_2 in a neighborhood of w_0. It follows from the analyticity of the arcs emanating from z_0 and equation (2.1.2) that (2.1.3) is uniformly bounded from zero; hence $d(z, C_\rho) \geqq K(\rho - 1)^t$ for z on C and in a sufficiently small neighborhood of the corner z_0; the constant $K > 0$ depends upon the neighborhood, but is independent of ρ for ρ near 1. Since this inequality holds in a neighborhood of each corner, and since the arcs joining the corners are analytic we obtain a lower bound for $d(C, C_\rho)$;

---

* See also Warschawski [1934a] p. 324; Carathéodory [1932].

let the corners be designated in such a way that $2 \geqq \mu_1 \geqq \mu_2 \geqq \cdots \geqq \mu_\lambda$, and let $t = \mu_1$ if $\mu_1 > 1$ and $t = 1$ if $\mu_1 \leqq 1$, then $d(C,C_\rho) \geqq K(C)(\rho-1)^t$. If the curve C has a cusp we have $t = \mu_1 = 2$.

It is to be observed that in our evaluation of $d(z,C_\rho)$, z on C, the exponent t as well as the constant K depends on z. Let C be an arbitrary Jordan curve which contains an analytic Jordan arc γ, then $d(z,C_\rho) \geqq K(\rho-1)$ for z on a closed sub arc ω of γ containing neither end-point of γ; the constant K depends on ω.

Let C be the segment $-1 \leqq z \leqq 1$, then by inspection of the mapping function we see that $d(C,C_\rho) \geqq K(\rho-1)^2$; here it is not possible to take $t < 2$ due to the behavior of the mapping function at the end-points of the segment. However we have $d(z,C_\rho) \geqq K_1(\rho-1)$ for $-1 < a \leqq z \leqq b < 1$, where the constant K_1 depends on a and b.

In the following definition we make the exponent t the basis of an important classification:

DEFINITION 2.1.1. Let C be a Jordan arc or curve. We say that C is of Type t provided that (i) on C the chord and arc are infinitesimals of the same order, and (ii) we have $d(C,C_\rho) \geqq K(C)(\rho-1)^t$, where K(C) is a positive constant depending on C but independent of ρ.

The requirement that for a curve or arc of Type t the chord and arc be infinitesimals of the same order is inserted to simplify the exposition in connection with approximation.

We see from the above discussion of curves with corners that the exponent t in the evaluation of $d(z,C_\rho)$ may have any positive value not greater than 2. In order to establish an upper bound on t for every C we prove the following:

THEOREM 2.1.2. Let R be a simply connected

region in the z-plane containing the point $z = 0$. Let $z = \Psi(w)$ map R conformally on $|w| < 1$ so that $z = 0$ corresponds to $w = 0$. Then we have

$$(2.1.4) \quad \liminf_{w \to e^{i\theta}} |\Psi(w) - \Psi(w_o)| \geq \frac{a}{16}(1-r)^2,$$

where $0 < r < 1$, $|w_o| = r$, and $|\Psi'(0)| = a$.

We use the Verzerrungssatz in the proof of this theorem. The function

$$\zeta = \frac{w - w_o}{1 - \bar{w}_o w}$$

maps $|w| < 1$ on $|\zeta| < 1$ so that the point $w = w_o$ corresponds to the point $\zeta = 0$; hence the function $g(\zeta) = \Psi(w)$ is schlicht in $|\zeta| < 1$. Furthermore $g(0) = \Psi(w_o)$ and

$$g'(\zeta) = \Psi'(w)\frac{dw}{d\zeta} = \frac{\Psi'(w)(1-\bar{w}_o w)^2}{1 - |w_o|^2},$$

whence $g'(0) = \Psi'(w_o)(1-|w_o|^2)$. Hence by the Verzerrungssatz for the modulus of the mapping function we have

$$\left| \frac{g(\zeta) - g(0)}{\Psi(w_o)-(1-|w_o|^2)} \right| \geq \frac{|\zeta|}{(1+|\zeta|)^2},$$

or

$$|\Psi(w)-\Psi(w_o)| \geq |\Psi'(w_o)|(1-|w_o|^2)\frac{|w-w_o||1-\bar{w}_o w|^2}{|1-\bar{w}_o w|(|1-w_o w|+|w-w_o|)^2}$$

But by the Verzerrungssatz for the derivative of the mapping function we have

$$|\Psi'(w_0)| \geq \frac{1-|w_0|}{(1+|w_0|)^3} \; |\Psi'(0)|,$$

hence

$$|\Psi(w)-\Psi(w_0)| \geq |\Psi'(0)| \frac{(1-|w_0|)^2 |w-w_0| \, |1-\bar{w}_0 w| (1+|w_0|)}{(1+|w_0|)^3 (|1-\bar{w}_0 w|+|w-w_0|)^2},$$

whence (2.1.4) follows by letting w approach $e^{i\theta}$. We use in (2.1.4) the lower limit since a unique limit may not exist.

Now let C be a Jordan arc or curve and let $z = \psi(w)$ be the usual mapping function; the reciprocal transformation and Theorem 2.1.2 yields

$$|\psi(w_1) - \psi(w_2)| \geq K(C)\cdot(\rho-1)^2,$$

where $|w_1| = \rho > 1$, $|w_2| = 1$, and K(C) is a positive constant depending on C but independent of ρ. Thus we see that for z an arbitrary point on C $d(z,C_\rho) \geq K(\rho-1)^2$, from which we conclude that $d(C,C_\rho) \geq K(C)\cdot(\rho-1)^2$. Consequently every Jordan arc or curve whose chord and arc are infinitesimals of the same order is an arc or curve of Type 2. However, as we have seen there are many Jordan curves not merely of Type 2 but of Type t where $t < 2$. An analytic Jordan curve is of Type 1. If the curve with corners discussed above has no cusp, that is if $\mu_1 \neq 2$, then the curve is of Type t, $t < 2$; if the curve has a cusp then the chord and arc are not infinitesimals of the same order and hence the curve is not included in our classification. Every Jordan arc whose chord and arc are infinitesimals of the same order is an arc of Type 2.

We are now ready to begin our study of polynomials. Before proceeding to the estimation of the derivative we prove a theorem which is of fundamental importance later as well as in the present section.

THEOREM 2.1.3. Let E with boundary C be a closed limited point set whose complement is connected and regular. Suppose $|P_n(z)| \leqq M$, z on C, where $P_n(z)$ is an arbitrary polynomial of degree n and M is a constant. Then $|P_n(z)| \leqq M\rho^n$, z on or within C_ρ.

We form the function $P_n(z)/[\phi(z)]^n$, where $\phi(z)$ is the usual (§1.1) mapping function. This function is analytic exterior to E, except for branch points, even at infinity if suitably defined there; the function itself is not necessarily single valued but its modulus is single valued exterior to E. By the principle of the maximum its modulus cannot attain a maximum exterior to E; we exclude the case of a constant for which the theorem is obvious. As z approaches C the modulus of $\phi(z)$ approaches unity and hence the modulus of $P_n(z)/[\phi(z)]^n$ can approach no limit greater than M; consequently we have $|P_n(z)/[\phi(z)]^n| \leqq M$, z exterior to E. But on C_ρ the modulus of $\phi(z)$ is by definition ρ and hence we have $|P_n(z)| \leqq M^n_\rho$, z on C_ρ; moreover the function $P_n(z)$ is analytic on and within C_ρ. Thus we have merely to apply the principle of the maximum to complete the proof of the theorem.

Now let E with boundary C be a closed limited set whose complement is simply connected and let $|P_n(z)| \leqq M$, z on E. By the Cauchy integral formula we have

$$P_n'(z_0) = \frac{1}{2\pi i}\int_\sigma \frac{P_n(t)}{(t-z_0)^2}\,dt,$$

where z_0 is an arbitrary point of E, and σ is a circle with center at z_0. Let the radius of σ be $K(C)(\rho-1)^t$, where t and K(C) are defined in Definition 2.1.1. Then σ, including its boundary, lies in $\overline{C}_\rho$, hence by Theorem 2.1.3 we see that the function $|P_n(t)| \leqq M\rho^n$, t on σ. Thus we have

$$|P_n'(z_0)| \leqq \frac{1}{2\pi} \frac{M\rho^n 2\pi K(C)(\rho-1)^t}{[K(C)(\rho-1)^t]^2}, \quad \rho > 1,$$

$$\leqq \frac{M}{K(C)} \frac{\rho^n}{(\rho-1)^t};$$

putting $\rho = 1 + 1/n$ we obtain

$$\frac{\rho^n}{(\rho-1)^t} = \frac{(1+\frac{1}{n})^n}{(\frac{1}{n})^t} \leqq en^t.$$

Thus we have proved the following theorem:

THEOREM 2.1.4. Let C be a Jordan arc or curve of Type t and let $|P_n(z)| \leqq M$, z on C, where $P_n(z)$ is a polynomial of degree n and M is a constant. Then

$$|P_n'(z)| \leqq K(C)Mn^t, \quad z \text{ on } \overline{C},$$

where K(C) is a constant depending on C, but not on $P_n(z)$, n, and z.

Suppose E, not a single point, is a closed limited set whose complement is simply connected; let C denote the boundary of E. Then since $d(C, C_\rho) \geqq K(C)(\rho-1)^2$ we see by the proof of Theorem 2.1.4 that $|P_n(z)| \leqq M$, z on E, implies $|P_n'(z)| \leqq K(C)Mn^2$, z on E.

It is also important to note that if in Theorem 2.1.4 we have for a particular point z_0 of C the inequality $d(z_0, C_\rho) \geqq K(z_0)(\rho-1)^{t_0}$, then $|P_n'(z_0)| \leqq K(z_0)Mn^{t_0}$. Thus if $|P_n(z)| \leqq M$ for z on E: $-1 \leqq z \leqq 1$, we have $|P_n'(z)| \leqq K(E)Mn^2$ for z on E, but we have $|P_n'(z)| \leqq K(a,b)Mn$ for z on the segment $-1 < a \leqq z \leqq b < 1$.

Theorem 2.1.4 is fundamental in the study of Problem α. For C the unit circle the theorem follows immediately from a well known* result on the derivative of a trigonometric sum; here $t = 1$ and $K(C) = 1$. For E the line segment $-1 \leqq z \leqq +1$ the theorem is due to Markoff [1889] for polynomials with real coefficients and to M. Riesz [1914] in the general case; for this line segment $t = 2$ and $K(C) = 1$. Szegö [1925] studied the modulus of $P_n'(z)$ at critical points of a curve C, in particular at the corners. The theorem as stated is due to Sewell [1936, 1937]. Theorem 2.1.2 is due to Szegö [1928]. Theorem 2.1.3 is due to Bernstein for E a line segment and to Walsh [1935] as stated.

§2.2. A POLYNOMIAL AND ITS INTEGRAL. In §2.1 we studied certain properties of the derivative of a polynomial on various sets implied by a bound on the polynomial; in the present section we study certain properties of a polynomial implied by a bound on the integral of a power of the polynomial. The former results are useful in the study of approximation in the sense of Tchebycheff, the latter to approximation as measured by a line integral.

We consider first the unit circle γ : $|z| = 1$. Let

$$(2.2.1) \quad P_n(z) \equiv a_0 + a_1 z + \cdots + a_n z^n,$$

and suppose

$$(2.2.2) \quad \frac{1}{2\pi} \int_\gamma |P_n(z)|^2 |dz| \leqq L^2,$$

where L is a constant. By the orthogonality property of $1, z, z^2, \ldots, z^n$, on γ we have from (2.2.2)

$$(2.2.3) \quad \frac{1}{2\pi} \int_\gamma |P_n(z)|^2 |dz| = |a_0|^2 + |a_1|^2 + \cdots + |a_n|^2 \leqq L^2;$$

furthermore by the Schwarz inequality,

* Bernstein [1926, pp. 44-46] considered trigonometric sums with real coefficients; the general case is due to M. Riesz [1914].

$$(2.2.4) \quad \left|\sum_{m=0}^{n} a_m b_m\right|^2 \leqq \sum_{m=0}^{n} |a_m|^2 \sum_{m=0}^{n} |b_m|^2,$$

we have from (2.2.1) and (2.2.3)

$$|P_n(z)|^2 \leqq L^2(n+1), \quad |z| = 1.$$

Hence

$$(2.2.5) \quad |P_n(z)| \leqq L(n+1)^{\frac{1}{2}} \leqq L'L\sqrt{n}, \quad |z| \leqq 1,$$

where L' is a constant independent of n and $P_n(z)$. If we take a_m and b_m proportional and $\arg(a_m b_m)$ independent of m in (2.2.4) the equality sign is valid; these conditions are fulfilled by the polynomial $P_n(z) \equiv 1+z+z^2+ \dots +z^n$ at the point $z = 1$.

For the general case we prove the following theorem:

THEOREM 2.2.1. Let C be a Jordan arc or curve of Type t and let

$$(2.2.6) \quad \int_C |P_n(z)|^p \, |dz| \leqq L^p, \quad p > 0,$$

where $P_n(z)$ is an arbitrary polynomial of degree n, and L and p are constants. Then we have

$$(2.2.7) \quad |P_n(z)| \leqq L'L\, n^{t/p}, \quad z \text{ on } \overline{C},$$

where L' is a constant depending on C and p, but independent of $P_n(z)$, n, and z.

In the special case of the circle considered above the value of t is unity and we took p = 2, hence $n^{1/2}$ in inequality (2.2.5) is precisely $n^{t/p}$; it follows from the example given that Theorem 2.2.1 cannot be improved for t = 1, p = 2 as far as the exponent of n is concerned.

We proceed with the proof. Let

$$\mu_n = \max_{z \text{ on } C} |P_n(z)|,$$

then by Theorem 2.1.4 we know that

$$|P_n'(z)| \leqq \mu_n K(C) n^t, \quad z \text{ on } \overline{C}.$$

Let $|P_n(z_0)| = \mu_n$ where z_0 is a point of C, then

$$(2.2.8) \quad |P_n(z) - P_n(z_0)| = \left| \int_{z_0}^{z} P_n'(z)dz \right| \leqq \mu_n K(C) n^t \int_{z_0}^{z} |dz|,$$

where the integral is taken along C from z_0 to z. Let s be the arc of C consisting of the set of points ζ : ζ on C, with the property that

$$(2.2.9) \qquad \int_{z_0}^{\zeta} |dz| \leqq \frac{1}{2AK(C)n^t},$$

where A is a constant so chosen that the length of s is not less than $1/[AK(C)n^t]$. It is sufficient to choose the constant A such that $A \geqq 2/K(C) \cdot l$, where l is the length of C; thus A is independent of the point z_0 and hence of $P_n(z)$. From (2.2.8) and (2.2.9) we see that

$$|P_n(z)| \geqq \left(\frac{2A-1}{2A}\right) \mu_n, \quad z \text{ on } s.$$

Thus it follows that

$$\int_C |P_n(z)|^p \, |dz| \geqq \left(\frac{2A-1}{2A}\right)^p \mu_n^p \, \frac{1}{AK(C)n^t},$$

hence by the definition of μ_n and (2.2.6) we have (2.2.7).

It is clear that the proof goes through with t = 2 for C an arbitrary rectifiable Jordan arc or curve.

For p an integer the theorem may be proved by using the fact that

$$(2.2.10) \qquad \int_{z_0}^{z} [P_n(z)]^p \, dz,$$

where the path of integration is along C, is a polynomial of degree pn+1 which is bounded on C by L^p. The derivative of this polynomial is precisely $[P_n(z)]^p$; thus we obtain a bound on $P_n(z)$ itself by applying Theorem 2.1.4.

Theorem 2.2.1 for polynomials normal and orthogonal on C is due to Sewell [1937a]. The method has been used by Jackson [1930, 1930a, 1931] on several occasions in connection with polynomials of best approximation in the sense of least p-th powers.

§2.3. THE LAGRANGE-HERMITE INTERPOLATION FORMULA.

In the study of approximation valuable results are obtained by studying certain interpolating polynomials. A particularly useful formula in this connection is the Lagrange-Hermite interpolation formula.

Let E, with boundary C, be a closed limited set and let C consist of a finite number of mutually exterior rectifiable Jordan curves. Let $f(z)$ be analytic in C and continuous on E. If the points $z_1, z_2, \ldots, z_{n+1}$, not necessarily distinct, all lie interior to C the unique polynomial $p_n(z)$ of degree n in z which interpolates to $f(z)$ in the points z_j, $j = 1,2,\ldots, n+1$, is

$$(2.3.1)\quad p_n(z) = \frac{1}{2\pi i}\int_C \frac{[\omega_n(t)-\omega_n(z)]f(t)}{\omega_n(t)(t-z)}\,dt,$$

z interior to C,

where $\omega_n(z) \equiv (z-z_1)(z-z_2)\ldots(z-z_{n+1})$. A useful form of (2.3.1) which emphasizes the approximating feature of $p_n(z)$ is

$$(2.3.2)\quad f(z)- p_n(z) = \frac{1}{2\pi i}\int_C \frac{\omega_n(z)f(t)dt}{\omega_n(t)(t-z)},$$

z interior to C;

Equations (2.3.1) and (2.3.2) are clearly equivalent by virtue of Cauchy's integral formula. The proof of their validity is immediate; from equation (2.3.2) we see that $f(z)$ and $p_n(z)$ coincide in n+1 points z_j, not necessarily distinct, for the function

$$\frac{1}{2\pi i}\int_C \frac{f(t)}{\omega(t)(t-z)}\,dt$$

is analytic interior to C, hence analytic in each of the points z_j. The integrand in (2.3.1) has no singularity in z, for the numerator vanishes identically for t = z and hence is divisible by t-z. The present numerator in (2.3.1) is a polynomial in z of degree n+1, so the function $p_n(z)$ defined by (2.3.1) is a polynomial in z of degree n; the proof of (2.3.2) is complete. Moreover, since equation (2.3.1) is valid for z interior to C, that equation is also valid for all values of z; we naturally suppose the integrand defined for z = t by a limiting process as z approaches t.

§2.4. GEOMETRIC PROPERTIES OF EQUIPOTENTIAL CURVES. The equipotential curves C_ρ of §1.1 play an important role in our study of approximation. It is clear from the definition of Problem β that these curves are fundamental in an investigation of that problem. Also we found it necessary in §2.1 to consider in detail certain properties of these curves. Furthermore in every application of formula (2.3.2) we take the points of interpolation z_j on E and integrate over C_ρ. In the present section we go into further detail concerning the geometric properties of these curves.*

Let E, with boundary C, be a closed limited point set of the z(= x+iy)-plane whose complement K (with respect to the extended plane) is connected and regular in the sense that K possesses a Green's function G(x,y) with pole at infinity. Then the function $w = \phi(z) = e^{G+iH}$, where H is conjugate to G in K, maps K conformally but not necessarily uniformly onto the exterior of the unit circle γ in the w-plane so that the points at infinity in the

* See Walsh [1937] for further details and bibliography.

two planes correspond to each other; interior points of K correspond to exterior points of γ, and exterior points of γ correspond to interior points of K.

The complement K is regular if it is connected and of finite connectivity provided E is closed and limited and has no isolated points; we restrict our discussion to this case. If K is simply connected we have a one to one continuous correspondence between the points of K and the points exterior to γ. If K is multiply connected the function $w = \phi(z)$ cannot set up such a correspondence so that $w = \phi(z)$ cannot be single valued in K. The branch points of $\phi(z)$ can have no limit point interior to K. The modulus of $\phi(z)$ is e^G, which is single valued in K.

If K is simply connected the situation is relatively simple. We see from the conformal map that each equipotential locus C_ρ : $G(x,y) = \log \rho > 0$, or $|\phi(z)| = \rho > 1$, is the transform of the circle $|w| = \rho$, and hence is an analytic Jordan curve. Moreover, the locus C_ρ always lies interior to the locus $C_{\rho'}$, $\rho' > \rho$.

A critical point of $G(x,y)$ is a point (interior to K) where both first partial derivatives of $G(x,y)$ vanish, or a point where $\phi'(z)$ vanishes; it will be noted that the vanishing of the derivative of any particular branch of $\phi(z)$ at a point $z = z_0$ implies the vanishing of the derivative of every branch of $\phi(z)$ at $z = z_0$.

Through an arbitrary point (x_0, y_0) of K passes one and only one locus C_ρ, namely, the locus $G(x,y) = G(x_0, y_0)$. In the neighborhood of such a point, the locus C_ρ is the image of the circle $|w| = \rho$ under the transformation $w = \phi(z)$, and hence C_ρ consists of a single analytic arc if $\phi'(x_0+iy_0) \neq 0$ and of $m+1$ branches through (x_0, y_0) with equally spaced tangents if (x_0,y_0) is an m-fold root of $\phi'(z)$ (i.e., an m-fold critical point of $G(x,y)$). In the neighborhood of every point of C_ρ there are points (x,y) of K where $G(x,y) > \log_\rho$ and also points (x,y) where $G(x,y) < \log \rho$.

When (x,y) approaches C, the function $G(x,y)$ approaches zero; when (x,y) becomes infinite, the function $G(x,y)$ becomes infinite. Every Jordan arc joining C to infinity cuts each C_ρ. The locus C_ρ consists of a finite number of mutually exterior analytic Jordan curves unless critical points of $G(x,y)$ be on the locus. In the latter case the locus C_ρ consists of a finite number of Jordan curves which are mutually exterior except for such critical points. Thus C_ρ consists of a finite number of finite Jordan curves which have a totality of no more than a finite number of intersections; each Jordan curve is composed of a finite number of analytic Jordan arcs. Every point of C lies interior to one and only one such Jordan curve of a given C_ρ.

It follows from the monotonic character of the loci that when ρ is sufficiently small the locus C_ρ consists of one contour surrounding each component of C; furthermore C_ρ is always interior to $C_{\rho'}$ if $\rho < \rho'$. If K is of connectivity q, the point at infinity considered an interior point, then no C_ρ can consist of more than q curves.

If the function $w = \phi(z)$ maps K onto the exterior of γ so that the points at infinity in the two planes correspond to each other, the function $w = \phi(z)/\rho$ similarly maps the exterior of C_ρ onto the exterior of γ. That is to say, the locus $[C_\rho]_{\rho'}$ is the locus $|\phi(z)/\rho| = \rho'$, or the locus $C_{\rho\rho'}$.

Let E', with boundary C', be a proper subset of E and let $G'(x,y)$ be Green's function with pole at infinity for the complement K' (supposed connected and regular) of E'. Let C'_ρ, denote the locus $G'(x,y) = \log \rho > 0$. Then C'_ρ lies interior to C_ρ.

A particular set whose equipotential loci can be readily studied is one bounded by a lemniscate. A _lemniscate_ Γ_μ is defined as the locus of points z of the form

$$(2.4.1)\quad \Gamma_\mu : |p(z)| = \mu > 0,\ p(z) \equiv (z-\beta_1)(z-\beta_2)\cdots(z-\beta_\lambda),$$

the numbers μ and the points β_j, $j = 1,2,\ldots,\lambda$, being fixed; the points β_j are not necessarily all distinct. A lemniscate then consists of a finite number of Jordan curves which are mutually exterior except possibly for a finite number of points each of which may belong to several of the Jordan curves; these Jordan curves are analytic except at such common points.* The function $\phi(z)$ can be chosen as $[p(z)/\mu]^{1/\lambda}$, and the locus C_ρ is the lemniscate $|p(z)| = \mu\rho^\lambda$.

As far as interpolation is concerned the formula (2.3.2) yields some immediate results for the lemniscate. It is natural to choose as the points of interpolation the points β_j of the lemniscate Γ_μ. Let $f(z)$ be analytic in the interior of the lemniscate Γ_{μ_1}, $\mu_1 > \mu$, and continuous on $\overline{\Gamma}_\mu$, and let us consider the polynomial $p_n(z)$ of degree $n = m\lambda - 1$ which interpolates to $f(z)$ in each of the points β_j counted m times. In this case we have $\omega_n(z) \equiv [p(z)]^m$, whence

$$(2.4.2) \quad \left| \frac{\omega_n(z)}{\omega_n(t)} \right| = \left(\frac{\mu}{\mu_1} \right)^m, \quad z \text{ on } \Gamma_\mu, \ t \text{ on } \Gamma_{\mu_1},$$

which through formula (2.3.2) leads directly to results on degree of approximation. If we denote Γ_μ by C so that C_ρ is the locus $|p(t)| = \mu\rho^\lambda$, equation (2.4.2) may be written

$$(2.4.3) \quad \left| \frac{\omega_n(z)}{\omega_n(t)} \right| = \frac{1}{\rho^{m\lambda}} = \frac{1}{\rho^{n+1}}, \quad z \text{ on } C, \ t \text{ on } C_\rho .$$

In order to study the situation for an arbitrary n it is convenient to set $n+1 = q\lambda + r$, $0 \leq r < \lambda$, and to take as the n+1 points of interpolation the points β_j, $j = 1,2,\ldots,\lambda$, each counted q times and in addition the points $\beta_1, \beta_2, \ldots, \beta_r$. Then we have for z on Γ_μ

---

* For a more detailed discussion of lemniscates see, e. g., Montel [1910]; Walsh [1935], pp. 54-56.

and t on Γ_{μ_1}

$$\left|\frac{\omega_n(z)}{\omega_n(t)}\right| = \left(\frac{\mu}{\mu_1}\right)^q \left|\frac{(z-\beta_1)\cdots(z-\beta_r)}{(t-\beta_1)\cdots(t-\beta_r)}\right| .$$

The second factor of the right member is uniformly bounded, so we have

$$(2.4.4) \quad \left|\frac{\omega_n(z)}{\omega_n(t)}\right| \leq M \left(\frac{\mu}{\mu_1}\right)^q \leq M_1 \left(\frac{\mu}{\mu_1}\right)^{n/\lambda} ,$$

uniformly for z on Γ_μ, t on Γ_{μ_1}; thus we have uniformly

$$(2.4.5) \quad \left|\frac{\omega_n(z)}{\omega_n(t)}\right| \leq \frac{M}{\rho^n} , \ z \text{ on } C, \ t \text{ on } C_\rho ,$$

where C denotes Γ_μ and C_ρ denotes Γ_{μ_1}.

§2.5. EQUALLY DISTRIBUTED POINTS. In formula (2.3.2) it is clear that a suitable choice of points of interpolation yields results on degree of approximation of the corresponding interpolating polynomial. For the lemniscate we saw in the preceding section that choosing the poles of the lemniscate as points of interpolation leads immediately to an inequality on $|\omega_n(z)/\omega_n(t)|$; however a lemniscate is a restricted type of point set. In the present section we consider a set bounded by a finite number of mutually exterior Jordan curves; our principal goal is to obtain an inequality on $|\omega_n(z)/\omega_n(t)|$; for z on C and t on C_ρ, for points of interpolation z_j, j = 1.2. ..., n+1, suitably chosen on C.

Since the unit circle is a lemniscate with a single pole at the origin it is clear from (2.4.5) that if we take the points of interpolation z_{j_n} coinciding at the origin we have $|\omega_n(z)/\omega_n(t)| \leq M/\rho^n$, uniformly for z on C; $|z| = 1$, and t on C_ρ : $|z| = \rho > 1$; in fact we have $|\omega_n(z)/\omega_n(t)| = 1/\rho^n$. This remark has the disadvantage of not extending at once from the unit circle to a set of more general Jordan curves. However if we take as the n+1 points of interpolation z_j the (n+1)-th roots of unity, all on $|z| = 1$, we have $\omega_n(z) = z^{n+1} - 1$, whence $|\omega_n(z)/\omega_n(t)| \leq M/\rho^n$, uniformly for z on C and t on C_ρ; the constant M is independent of n, z, and t. This set of points does admit of generalization to points belonging to a set of Jordan curves. In fact the present section is devoted to the study of $|\omega_n(z)/\omega_n(t)|$, z on C and t on C_ρ, where the points z_j are the transforms of the roots of unity under the conformal map of the complement of E (notation of §1.1) onto $|w| > 1$; these are called equally distributed points* and the method is classical**. Let E, with boundary C, be a closed limited set bounded by a finite number of mutually exterior Jordan curves. If C is a single Jordan curve the correspondence is one to one, and its mapping function and its inverse single valued, hence the situation is relatively easy to handle; for several curves the definition of equally distributed points requires some explanation.

Let us assume for definiteness and brevity that C consists of two mutually exterior rectifiable Jordan curves C_1 and C_2, whose continuity properties are to be more precisely described later; the extension to any finite number of curves involves only slight modification. Let K be the complement of E with respect to the extended plane, then the Green's function for K may be written in

* The term is due Féjer [1918].

** See Walsh [1935], pp. 65-75. The method is due to Hilbert, Féjer, and others. We follow closely the exposition in Walsh and Sewell [1940].

the form

$$G(x,y) = \log (x^2+y^2)^{1/2} + G_0(x,y),$$

where $G_0(x,y)$ approaches a constant $-g$ as (x,y) becomes infinite. If we set

$$(2.5.1) \quad V(x,y) \equiv G(x,y) + g$$

then under the assumption that $\partial V/\partial \nu$, where ν is the exterior normal for C, is uniformly bounded and measurable on C, we have

$$(2.5.2) \quad V(x,y) = \int_C \phi_1(s) \log r \, ds, \quad \phi_1(s) = \frac{1}{2\pi} \frac{\partial V}{\partial \nu},$$

where $r = |z-\zeta|$, $ds = |d\zeta|$, and $z = x+iy$ is an arbitrary point of K. Formula (2.5.2) is easy to prove. Let C' denote a circle whose center is $P:(x,y)$ which contains C in its interior. We have

$$G(x,y) = \frac{1}{2\pi} \int_C \left(\log r \cdot \frac{\partial G}{\partial \nu} - G \frac{\partial \log r}{\partial \nu}\right) ds$$

$$(2.5.3) \qquad + \frac{1}{2\pi} \int_C \left(\log r \cdot \frac{\partial G}{\partial \nu} - G \frac{\partial \log r}{\partial \nu}\right) ds$$

where as before $r = |z-\zeta|$, $ds = |d\zeta|$, and ν is the interior normal for the region bounded by C' and C; equation (2.5.3) is valid even if C' depends on (x,y). In the second integral on the right-hand side we make the substitution $G = \log r + G_1$. Even though r is not $|\zeta|$, the function $G_1(\xi, \eta)$ approaches the value $-g$ as $\zeta = \xi + i\eta$ becomes infinite. By Gauss's mean value theorem applied to the exterior of C' we have

$$g = \frac{1}{2\pi} \int_{C'} G_0 \frac{\partial \log r}{\partial \nu} ds = \frac{1}{2\pi} \int_{C'} G \frac{\partial \log r}{\partial \nu} ds,$$

and we have also

$$\int_{C'} \log r \frac{\partial G_1}{\partial \nu} ds = 0.$$

If we use the fact that G vanishes on C, we now have

$$G(x,y) + g = \frac{1}{2\pi} \int_C \log r \cdot \frac{\partial (G+g)}{\partial \nu} \, ds,$$

which is equivalent to (2.5.2).

Let

$$(2.5.4) \qquad u_o = \int_C \phi_1(s)\, ds, \quad u(\zeta) = \int_o^{s(\zeta)} \phi_1(s)\, ds \; ;$$

here we know that

$$(2.5.5) \qquad u_o = \int_C \phi_1(s)\, ds = \frac{1}{2\pi} \int \frac{\partial V}{\partial \nu} ds = - \frac{1}{2\pi} \int_C \frac{\partial \log r}{\partial \nu} ds = 1.$$

Also by §2.4 the function $\phi_1(s)$ is positive on C, hence

$$\int_{C_1} du = \lambda \quad , \quad \int_{C_2} du = 1 - \lambda \; , \; 0 < \lambda < 1.$$

The function $w = \phi(z)$ which maps K onto $|w| > 1$ is not uniquely determined, and even when a definite choice is made is multiple valued. If we choose a particular branch, the function sets up a correspondence between C_1 and an arc of length $2\pi\lambda$ of the circle $\gamma : |w| = 1$. This arc may begin at any arbitrary point on γ and an end-point of this arc may correspond to any preassigned point of C_1; a similar correspondence is set up between C_2 and an arc of length $2\pi(1-\lambda)$ on γ. In order to obtain a one to one correspondence which is complete we take each of the two arcs of γ as closed at one end and open at the other; then we preassign a fixed point of C_1, a fixed point of γ, and a fixed point of C_2. If we start with this fixed point of C_1 and fixed point of γ we obtain a one to one correspondence between C_1 and an arc β on γ of length $2\pi\lambda$ measured from the preassigned point of γ; also we set the other end-point of β in correspondence to the preassigned point of C_2, with the correspondence continuing around C_2 and around the circle back to the original fixed point of γ. Thus we have a one to one correspondence between C and γ, a correspondence which is continuous on C except at a single point of C_1 and a single point of C_2, and which is likewise continuous on γ except at points of γ corresponding to the two points of C_1 and C_2 respectively. This correspondence, once chosen, is to be temporarily fixed during the subsequent discussion; we shall eventually need another similar correspondence. The continuity and differentiability properties of the correspondence are determined, except for the two discontinuities, by the continuity and differentiability properties of C_1 and C_2 alone. We are interested in these properties merely in the closed exterior neighborhoods of C and γ; for our purposes we do not need continuity properties remote from C and γ.

With the correspondence defined we consider more closely the equally distributed points. From (2.5.2), (2.5.4), and (2.5.5) we have

$$(2.5.6) \qquad V(x,y) = \int_0^1 \log r \, du, \quad z \text{ in } K,$$

and consequently by definition of the definite integral we have for z in K

$$(2.5.7) \qquad V(x,y) = \lim_{n \to \infty} \frac{1}{n}(\log r_1 + \log r_2 + \dots + \log r_n),$$

where $r_1, r_2, \dots, r_n$ are the distances from z to the n points ζ_m of C which correspond to the n equidistant values k/n, k = 1,2,...,n of u in the interval (0,1). The points ζ_m depend of course upon n, but for simplicity we omit that dependence in the notation. The points ζ_m may be considered as the transforms of the roots of unity on γ, and hence we have a method whereby the equally distributed points, referred to above, are determined.

Our next step is to investigate the behavior of

$$(2.5.8) \qquad \omega_{n-1}(z) \equiv (z-\zeta_1)(z-\zeta_2) \dots (z-\zeta_n),$$

both for z on C and for z in K. Equations (2.5.6) and (2.5.7) furnish formulas which serve to study $\omega_{n-1}(z)$ for z in K. In fact (2.5.6) is valid also for z on C under certain restrictions on C as the following lemma shows:

LEMMA 2.5.1. Let C consist of the mutually exterior rectifiable Jordan curves $C_1, C_2, \ldots, C_\lambda$ and let C have a tangent at every point. Let $\partial G/\partial\nu$ be uniformly bounded and measurable on C, and let a constant A exist such that

$$(2.5.9) \qquad \left| \int_{\zeta_1}^{\zeta_2} ds \right| \leqq A |\zeta_1 - \zeta_2|,$$

for arbitrary ζ_1 and ζ_2 on a particular C_j, $j = 1, 2, \ldots, \lambda$, where the integral is taken over C along the shorter of the two paths joining ζ_1 to ζ_2. Then the equation (2.5.2) is valid for z on C as well as in K.

It is to be observed that equations (2.5.2) and (2.5.6) are equivalent by virtue of (2.5.4) and (2.5.5). Consequently by the lemma we have the validity of (2.5.6) on C and hence an approach to the investigation of $\omega_{n-1}(z)$ for z on C.

We proceed with the proof. Let $z_0 = x_0+iy_0$ be a fixed point on C. Since V(x,y), when suitably defined on C, is known to be continuous on the closure of K, it is sufficient for us to prove that

$$(2.5.10) \qquad \lim_{z_1 \to z_0} \int_C \frac{\partial V}{\partial \nu} \log \left| \frac{z_1 - \zeta}{z_0 - \zeta} \right| ds = 0,$$

where z_1 is in K and lies on the normal to C through z_0. The existence of the integral in (2.5.10) will appear in the course of the proof. Let $\eta > 0$ be arbitrary with the condition that $\eta < 1$ and $\log(1-\eta^2) > -\eta$, from which it follows that $\log(1+\eta^2) < \eta$. Let Ω_1 be the set of points ζ on C: $|\zeta - z_0| \geqq \eta$, and let Ω_2 be the set $C - \Omega_1$. Let z_1 be a point in K such that $|z_1 - z_0| < \eta^3$, then we have for ζ on Ω_1

$$\left|\frac{z_1 - \zeta}{z_0 - \zeta}\right| \leqq \frac{|z_1 - z_0| + |z_0 - \zeta|}{|z_0 - \zeta|} \leqq 1 + \eta^2,$$

$$\left|\frac{z_1 - \zeta}{z_0 - \zeta}\right| \geqq \frac{|z_0 - \zeta| - |z_1 - z_0|}{|z_0 - \zeta|} \geqq 1 - \eta^2,$$

$$\left|\log\left|\frac{z_1 - \zeta}{z_0 - \zeta}\right|\right| < \eta, \quad \zeta \text{ on } \Omega_1.$$

Since $|\partial G/\partial \nu| \leqq M$ we have

$$(2.5.11) \qquad \left|\int_{\Omega_1} \frac{\partial V}{\partial \nu} \log\left|\frac{z_1 - \zeta}{z_0 - \zeta}\right| ds\right| \leqq l\, M\eta,$$

where l is the length of C.

For ζ on Ω_2 we have

$$|\zeta - z_1| \leqq |\zeta - z_0| + |z_0 - z_1| \leqq \eta + \eta^3,$$

$$\log\left|\frac{\zeta - z_1}{\zeta - z_0}\right| = \log|\zeta - z_1| - \log|\zeta - z_0|$$

$$\leqq \log(\eta + \eta^3) - \log|\zeta - z_0|.$$

Let $\alpha < \pi/2$ be a number such that Ω_2 lies in a closed double sector of angle 2α with vertex z_0; since C has a tangent at every point such a number obviously exists. As a matter of fact with z_1 on the normal to C at z_0 we can choose α so that (with possibly a further restriction on η) z_1 lies on the bisector of the exterior angle of this sector. The distance from ζ on Ω_2 to the bisector is not greater than $|\zeta - z_1|$, and is at least as great as $|\zeta - z_0| \cos\alpha$, hence

$$|\zeta - z_1| \geq |\zeta - z_0| \cos\alpha ,$$

$$\log \left| \frac{\zeta - z_1}{\zeta - z_0} \right| \geq \log \cos\alpha , \qquad \zeta \text{ on } \Omega_2 .$$

Thus we have for ζ on Ω_2

$$\left| \log \left| \frac{\zeta - z_1}{\zeta - z_0} \right| \right| \leq - \log \cos\alpha + \log(\eta + \eta^3) - \log|\zeta - z_0| .$$

By (2.5.9) the path of integration is not greater than $2A\eta$, hence

$$\left| \int_{\Omega_2} \frac{\partial V}{\partial \nu} \log \left| \frac{z_1 - \zeta}{z_0 - \zeta} \right| ds \right| \leq 2\, MA \cdot \eta \log \cos\alpha$$

$$+ 2\, M A \eta \log \eta$$

$$(2.5.12) \qquad + 2MA \cdot \eta^2 - 2MA \int_{z_0}^{z_0 + \eta} \log|\zeta - z_0|\, |d\zeta| ,$$

where the last integral may be interpreted as an ordinary definite integral rather than a line integral since C has a tangent at every point and since (2.5.9) holds. Inequality (2.5.12) implies the existence of the second member of (2.5.2) for z on C. The right hand side of (2.5.12) becomes

$$(2.5.13) \quad -2MA\eta \log \cos\alpha + 2\, MA\eta \log\eta + 2MA\, \eta^2$$

$$+ 2MA\eta\,(1 - \log\eta) ,$$

which can be made smaller than any preassigned $\delta > 0$ by a suitable choice of $\eta > 0$.

Consequently it follows that given any preassigned $\epsilon > 0$ we can by virtue of (2.5.11), (2.5.12), and (2.5.13) choose η, $0 < \eta < 1$, $\log(1 - \eta^2) > -\eta$, with the double sector at z_0 having the prescribed properties, so that the right hand side of (2.5.11) plus (2.5.13) is less than ϵ. Thus we have shown that for z_1 in K on the normal to C at z_0 and $|z_1 - z_0| < \eta^3$, the integral of (2.5.10) is less than ϵ and the proof of the lemma is complete.

In the lemma we assumed conditions on $\partial G/\partial \nu$ on C; it is well to observe that the behavior of $\partial G/\partial \nu$ on C_j, one of the components of C, is entirely similar to the behavior of $\partial G_j/\partial \nu$, where G_j denotes the Green's function for the exterior of C_j with pole at infinity. If we map successively on the exterior of the unit circle so that the points at infinity in the two planes correspond to each other the exterior of C_1, the exterior of the transform of C_2, the exterior of the last transform of C_3, and so on, the last map is a transformation of the exterior of $\overline{C}$ on a region bounded by a finite number of analytic Jordan curves; in the latter situation $\partial G/\partial \nu$ is analytic in the transform of C.

We proceed to study $\omega_{n-1}(z)$ for z on C since we know now by the lemma that

$$(2.5.14) \qquad \int_0^1 \log r \, du = V(x,y) = g, \quad z \text{ on } C,$$

where $r = |z - \zeta|$, and where $\zeta = \zeta(u)$ is the running variable on C for the integration. If we let $z = \psi(w)$ be the inverse of $w = \phi(z)$, then (2.5.14) becomes

$$(2.5.14') \qquad \int_{|w'|=1} \log |\psi(w) - \psi(w')| \, du = g,$$

where w lies on the circle $|w| = 1$. Equations (2.5.14) and (2.5.14') are <u>to be considered identical</u>; a consequence of this convention is that in (2.5.14') we assume a one to one correspondence between the points of $|w'| = 1$ and the points of C. This correspondence is, as we have already said, not uniquely determined, and once it has been determined is not single valued (possibly infinitely many valued) on $|w| = 1$. We do not assume any necessary connection between the various branches or determinations of $\psi(w)$ employed on the various arcs γ_j, provided merely that the correspondence between points of $|w'| = 1$ and points of C is continuous on each such arc and in the large is one to one.

In equations (2.5.14) and (2.5.14') we have $w' = \phi(\zeta) = e^{V+iV_1-g}$ where V_1 is conjugate to V in K; thus for ζ on C and $|w'| = 1$ we have $dw' = w'(dV+idV_1)$, $|dw'| = dV_1 = (\partial V/\partial \nu)ds = 2\pi du$, by virtue of the fact that V is constant on C.

A special case of (2.5.14), valid by Lemma 2.5.1, occurs when C is the unit circle $|w| = 1$:

$$(2.5.15) \qquad \frac{1}{2\pi} \cdot \int_{|w'|=1} \log |w - w'| \, |dw'| = 0,$$

for w on the circle $|w| = 1$. Through the transformation of the original region K onto $|w| > 1$ equations (2.5.14) and (2.5.15) can be combined to yield the equation

$$(2.5.16) \qquad \int_{|w'|=1} \log \left| \frac{\psi(w) - \psi(w')}{w - w'} \right| du = g$$

where w lies on the circle $|w| = 1$. We have thus replaced a difference by a difference quotient, which is a distinct advantage, especially since we are dealing with logarithms.

We are going to approximate the definite integral (2.5.16) by a sequence of Riemann sums (see (2.5.7)); it is well known that such sums approach the limit at a rate dependent upon the continuity properties of the integrand. In fact our method of evaluating $\omega_{n-1}(z)$ consists in approximating (2.5.16) by sums formed from equally distributed points; we assume sufficient continuity of C to ensure the desired degree of approximation of the integral by the Riemann sums. Let

$$(2.5.17) \qquad \tau_n = \frac{\omega_{n-1}(z)}{w^n - 1} =$$

$$\frac{[\psi(w) - \psi(w_1)][\psi(w) - \psi(w_2)] \ldots [\psi(w) - \psi(w_n)]}{(w-w_1)(w-w_2) \ldots (w-w_n)},$$

where $\omega_{n-1}(z) \equiv (z-\zeta_1)(z-\zeta_2) \ldots (z-\zeta_n)$, $\zeta_m = \psi(w_m)$, $w_m = e^{2\pi i\, m/n}$, $m = 1, 2, \ldots, n$. Then $\log \tau_n$ is a Riemann sum corresponding to the integrand of (2.5.16) with the points of subdivision in the roots of unity on $|w| = 1$. It is clear that an evaluation of $\log \tau_n$ leads directly to an evaluation of $\omega_{n-1}(z)$. Thus the following definition is appropriate:

DEFINITION 2.5.2. The set C of mutually exterior Jordan curves $C_1, C_2, \ldots, C_\lambda$ is called a contour provided: (i) each C_j has a tangent at every point; (ii) the function

$$(2.5.18) \qquad \log \left| \frac{\psi(w) - \psi(w')}{w - w'} \right|$$

for each possible definition of $\psi(w)$ and $\psi(w')$ is locally bounded in the two-dimensional sense for $|w| = 1$ and $|w'| \geqq 1$ in the neighborhood of each point of continuity of $\psi(w)$; (iii) the function (2.5.18) satisfies on each arc γ_j of γ corresponding to a curve C_j of C a Lipschitz condition of order unity in w, uniformly with respect to w' on any closed arc of $|w'| = 1$ interior to γ_k.

On each arc γ_j of γ corresponding to a curve C_j a single branch of $\psi(w)$ is to be used in connection with this definition.

The interpretation and significance of this definition requires some explanation. Each arc γ_j is closed at one end, open at the other end; such end-points are points of discontinuity of $\psi(w)$, for on one side of such a point the image point $z = \psi(w)$ lies on one C_j and on the other side of this end-point the image point lies on another C_j. This unusual and purely accidental character disappears if the interpretation is geometric in the light of the requirement that the Lipschitz condition shall be satisfied no matter how $\psi(w)$ is defined, that is to say no matter which point of each C_j is chosen to correspond to a γ_j, and no matter where γ_j lies on γ.

By definition the Lipschitz condition implies

$$(2.5.19) \quad \left| \log\left| \frac{\psi(w_1) - \psi(w')}{w_1 - w'} \right| - \log\left| \frac{\psi(w_2) - \psi(w')}{w_2 - w'} \right| \right|$$

$$\leq L\,|w_1 - w_2|,$$

where L is independent of w_1, w_2, and w'. If we choose w_1 as that end-point of γ_1 which belongs to γ_1, w_2 as a fixed point interior to γ_1, and let w' vary on the other arc γ_j having w_1 as an end-point (not belonging to it, of course), it is clear that when w' approaches w_1 the distance $|\psi(w_1) - \psi(w_1)|$ is the distance from a point of C_1 to a point of C_j and hence is bounded from zero, consequently the expression

$$\left| \frac{\psi(w) - \psi(w')}{w - w'} \right|$$

becomes infinite. When w' approaches w_1 the expression

$$\left| \frac{\psi(w_2) - \psi(w')}{w_2 - w'} \right|.$$

is bounded, so the left hand member of (2.5.19) becomes infinite and the inequality cannot hold uniformly with respect to w' with w' restricted merely by the equation $|w'| = 1$; this is our justification for requiring in Definition 2.5.2 uniformity with respect to w' merely for w' on any closed arc of $|w'| = 1$ interior to a γ_k. But we suppose here too that this condition holds for every choice of a point of C_k corresponding to an end-point of γ_k, that is to say, for an arbitrary choice of the correspondence between γ and C.

Actually the function (2.5.18) is not defined for $w = w'$. However, let us assume (2.5.19) valid, where w' as well as w_1 and w_2 lies interior to a particular arc γ_j. Let $W_1(w_1)$ denote any limiting value of the function

$$\log \left| \frac{\psi(w_1) - \psi(w')}{w_1 - w'} \right|$$

as w' approaches $w_1 \neq w_2$. By the uniformity of (2.5.19) and the continuity of $\psi(w)$ for $w_1 = w_2$ it follows that $W_1(w_1)$ is finite and

$$(2.5.20) \quad \left| W_1(w_1) - \log \left| \frac{\psi(w_2) - \psi(w_1)}{w_2 - w_1} \right| \right| \leqq L|w_1 - w_2|.$$

Now let w_2 approach w_1 and from (2.5.20) it follows that

$$\log \left| \frac{\psi(w_2) - \psi(w_1)}{w_2 - w_1} \right|$$

approaches the finite limit $W_1(w_1)$. Hence $W_1(w_1)$ is uniquely determined. By a change of notation in (2.5.20) we have

$$\left| W_1(w_2) - \log \left| \frac{\psi(w_1) - \psi(w_2)}{w_1 - w_2} \right| \right| \leqq L|w_1 - w_2|,$$

whence by (2.5.20) itself we have

$$(2.5.21) \quad |W_1(w_1) - W_1(w_2)| \leqq 2L|w_1 - w_2|,$$

that is, a Lipschitz condition on the function $W_1(w)$ on any arc γ_j.

When w' approaches w_1, both points lying on the same

γ_j, it is clear that

$$\arg \left[\frac{\psi(w_1) - \psi(w')}{w_1 - w'}\right]$$

approaches a limit since C_j has a tangent at every point. Hence as w' approaches w_1 the function

$$\log \left[\frac{\psi(w_1) - \psi(w')}{w_1 - w'}\right] =$$

(2.5.22)

$$\log \left|\frac{\psi(w_1) - \psi(w')}{w_1 - w'}\right| + i \arg \left[\frac{\psi(w_1) - \psi(w')}{w_1 - w'}\right]$$

approaches a limit, properly denoted by $\log \psi'(w)$, where the derivative is defined in the one-dimensional sense on γ_j, and is different from zero. But the function

$$\frac{\psi(w_1) - \psi(w')}{w_1 - w'},$$

considered as a function of w', with $|w'| \geq 1$ and w_1 a fixed interior point of γ_j, is continuous in the two-dimensional sense on any closed arc of $|w'| = 1$ interior to γ_j except perhaps at the point $w' = w_1$, it approaches the same limit when w' on γ_j approaches w in either sense, and by Definition 2.5.2 it is bounded in the two-dimensional sense in the neighborhood of w_1. It follows then from the discussion in connection with Theorem 1.2.10 that $\psi'(w_1)$ is actually a two-dimensional derivative and $\psi'(w_1) \neq 0$ by Definition 2.5.2.

The first member of (2.5.22), considered as a function of w', is continuous and its real part satisfies a Lipschitz condition of order unity on any closed arc of $|w'| = 1$ interior to γ_j. Privaloff [1916]* has shown that if the function $u(z)$ is harmonic in $|z| < 1$, is con-

* Or see, for instance, Zygmund [1935].

tinuous on $|z| \leq 1$, and satisfies a Lipschitz condition of order unity on $|z| = 1$ then the conjugate function $v(z)$ satisfies a Lipschitz condition of every order α, $0 < \alpha < 1$ on $|z| = 1$. In order to apply the result to the present situation we shall prove that

> If $u(z)$ is harmonic in the deleted neighborhood of a closed arc σ of the circumference $|z| = 1$, is continuous on σ in the two-dimensional sense, and satisfies a Lipschitz condition of order unity on σ, then the conjugate $v(z)$ defined on σ so as to be continuous there satisfies on any closed proper sub-arc of σ a Lipschitz condition of every order α, $0 < \alpha < 1$.

Let $u_1(z)$ be defined on $|z| = 1$ as equal to $u(z)$ on σ and satisfying a Lipschitz condition on the entire circumference (it is sufficient to take $u_1(z)$ exterior to σ as a linear function of arc length), and then let $u_1(z)$ defined by these boundary values be continuous for $|z| \leq 1$, harmonic for $|z| < 1$. By Privaloff's theorem the conjugate $v_1(z)$ of $u_1(z)$ satisfies a Lipschitz condition of every order α on $|z| = 1$. The function $u(z)-u_1(z)$ is harmonic in the deleted neighborhood of σ, is continuous on σ and equal to zero there, and can be extended harmonically across σ. The conjugate $v(z)-v_1(z)$ satisfies a Lipschitz condition of order α on any closed proper sub-arc of σ, hence so also does $v(z)$ itself. It follows from this result that the pure imaginary part of (2.5.22) (and therefore the function itself) satisfies on any closed arc of $|w'| = 1$ interior to γ_j a Lipschitz condition of every order α, $0 < \alpha < 1$. Thus by the reasoning used for inequality (2.5.19) in the proof of (2.5.20) we have on the arc considered

$$(2.5.23) \quad |\log \psi b(w_1) - \log \psi'(w_2)| \leq L|w_1 - w_2|^{\alpha},$$

for an arbitrary $\alpha < 1$, where L depends on α.

In our study of $\psi'(w)$ itself we need the following

LEMMA 2.5.3. Let $\log \Phi(w)$ satisfy a Lipschitz condition of given order α, $0 < \alpha \leqq 1$, on a closed proper sub-arc of $|w| = 1$. Then on that arc the function $\Phi(w)$ also satisfies such a condition.

Our hypothesis involves the boundedness of $\log \Phi(w)$ which is a consequence of the inequality

$$|\log \Phi(w_1) - \log \Phi(w_2)| \leqq L_1 \, |w_1 - w_2|^\alpha .$$

The continuity of $\Phi(w)$ on the arc considered follows from the continuity of $\log \Phi(w)$ there. For $|w_1 - w_2|$ sufficiently small to ensure that

$$|\Phi(w_1) - \Phi(w_2)| < \lambda \min |\Phi(w)|, \; 0 < \lambda < 1,$$

we have

$$\left|\log \frac{\Phi(w_1)}{\Phi(w_2)}\right| = \left| \left[\frac{\Phi(w_1)}{\Phi(w_2)} - 1\right] - \frac{1}{2}\left[\frac{\Phi(w_1)}{\Phi(w_2)} - 1\right]^2 + \cdots \right|$$

$$= \frac{|\Phi(w_1) - \Phi(w_2)|}{|\Phi(w_2)|} \left| 1 - \frac{1}{2}\left[\frac{\Phi(w_1)}{\Phi(w_2)} - 1\right] + \cdots \right| ;$$

this last factor is of the form

$$\left| \frac{\log \zeta}{\zeta - 1} \right|, \qquad |\zeta - 1| < \zeta < 1,$$

and hence is bounded from zero, whence by virtue of our hypothesis we have

$$(2.5.24) \qquad |\Phi(w_1) - \Phi(w_2)| \leqq L' \, |w_1 - w_2|^\alpha ,$$

for $|w_1 - w_2|$ sufficiently small. By the general

inequality*

$$| x_1 + x_2 |^p \leqq 2^{p-1} \; [| x_1|^p + | x_2|^p], \; p \geqq 1,$$

and by the use of (2.5.24) for two pairs of points (w_1, w_2) and (w_2, w_3), with w_2 between w_1 and w_3 we may write

$$|\Phi(w_1) - \Phi(w_3)|^{1/\alpha} \leqq 2^{(1-\alpha)/\alpha} \, [|\Phi(w_1) - \Phi(w_2)|^{1/\alpha}$$

$$+ |\Phi(w_2) - \Phi(w_3)|^{1/\alpha}]$$

$$\text{(2.5.25)} \qquad \leqq 2^{(1-\alpha)/\alpha} L^{1/\alpha} \, [|w_1 - w_2| + |w_2 - w_3|].$$

No matter what proper sub-arc of $|w| = 1$ is considered the ratio of the arc $w_1 \, w_2 \, w_3$ to the chord $w_1 \, w_3$ has a positive upper bound b which is independent of the points w_1, w_2, w_3; thus we have

$$|w_1 - w_2| + |w_2 - w_3| \leqq b \, |w_1 - w_3|,$$

whence by (2.5.25)

$$|\Phi(w_1) - \Phi(w_3)| \leqq L' \; |w_1 - w_2|^{\alpha} ,$$

where L' is independent of w_1, w_2 and w_3. By virtue of the uniform continuity of the function $\Phi(w)$ on the arc of $|w| = 1$ considered, it is now clear that use of (2.3.24) for a suitably chosen finite number of pairs of points w_1 and w_2 yields the conclusion of Lemma 2.5.3.

The restriction in the lemma that the closed arc considered should be a proper sub-arc of $|w| = 1$ is unnecessary, for if the arc is the entire circumference the proof just given requires only obvious modifications.

For the function $\psi'(w)$ on a contour we have

LEMMA 2.5.4. If C is a contour the function $\psi'(w)$ exists and is continuous in the

* See, e.g., Walsh [1935], p. 93.

two-dimensional sense on any closed arc interior to a γ_j, is different from zero on such an arc, and satisfies there a Lipschitz condition of every order α, $0 < \alpha < 1$.

We have already shown the existence in the two-dimensional sense of the derivative $\psi'(w)$ on the arc considered; the continuity in the one-dimensional sense of $\psi'(w)$ and Lipschitz condition of order α there are a consequence of (2.5.23) and Lemma 2.5.3; it follows from (2.5.23) that $\psi'(w) \neq 0$ on γ_j. The continuity of $\psi'(w)$ in the two-dimensional sense remains to be established. We might use the results of Theorems 1.2.12 and 1.2.14 by constructing a Jordan region whose boundary contains the arc, but a direct proof is easy. For z on C we have $ds = |dz|$, $z = \psi(w)$,

$$\frac{dz}{ds} = \frac{\psi'(w)}{|\psi'(w)|} \frac{dw}{|dw|} .$$

The functions $\psi'(w)$, $1/|\psi'(w)|$, $dw/|dw|$ all satisfy Lipschitz conditions of some order with respect to w, so dz/ds satisfies such a condition. We have also

$$|w_2 - w_1| \leq \int_{w_1}^{w_2} |dw| = \int_{s_1}^{s_2} \frac{ds}{|\psi'(w)|} \leq B|s_2 - s_1| ,$$

where B is a suitably chosen constant which is independent of w_1 and w_2, and where $w_1 = w(s_1)$, $w_2 = w(s_2)$. Consequently z(s) has the property that z'(s) exists and satisfies a Lipschitz condition of some order with respect to s. This property holds on every arc of C. It now follows from a theorem due to Kellogg [1912] that $\psi'(w)$ exists and is continuous in the two-dimensional sense on $|w| = 1$. Thus the proof of Lemma 2.5.4 is complete.

It follows from Lemma 2.5.4 that if C is a contour $\partial V/\partial \nu$ is uniformly bounded and measurable on C; moreover on C we have $z = \psi(w)$, $dz/ds = [\psi'(w)dw]/[|\psi'(w)dw|]$ so arc and chord are equivalent infinitesimals, thus (2.5.9) is valid for suitably chosen A. Consequently if C is a contour the hypothesis of Lemma 2.5.1 is fulfilled, a fact which enables us to use the formulas for z on C where C is a contour. We also conclude that each component of a contour is a curve of Type $t = 1$.

For the approximation of a definite integral by Riemann sums we need the following lemma [compare Walsh and Sewell, 1937]:

LEMMA 2.5.5. Let $f(x)$ be continuous in the closed interval $0 \leq x \leq a$ except for a finite number of finite discontinuities. Let $f(x)$ satisfy a Lipschitz condition in each subinterval of continuity. Then we have

$$\left| \int_b^c f(x)dx - \frac{a}{n} \sum f(am/n) \right| \leq M/n,$$

$$0 \leq b < c \leq a,$$

where M is a suitably chosen constant depending on the constant in the Lipschitz condition and on the sum of the discontinuities of $f(x)$ but not on n, b, or c, and where the summation is extended over the N points am/n in the interval $b \leq x < c$.

The proof is easy. We take $a = 1$ without loss of generality. Let $x_1, x_2, \ldots x_\lambda$ be the points of discontinuity of $f(x)$ and at each point x_j let the limits of $f(x)$ to the right and left be denoted respectively by $f(x_j^+)$ and $f(x_j^-)$ (of course $f(0^-)$ and $f(1^+)$ need not exist). If we set

$$S(x) = \sum_{0 < x_j \leq x} [f(x_j) - f(x_j^-)] + \sum_{0 \leq x_j < x} [f(x_j^+) - f(x_j)],$$

the function $f(x) - S(x)$ is continuous in the closed interval $0 \leq x \leq 1$ and satisfies there a Lipschitz condition of order unity. Let $F(x) = f(x) - S(x)$, then throughout the interval $(m-1)/n \leq x \leq m/n$ we have $|F(x) - F(m/n)| \leq L/n$, where L is the Lipschitz constant; hence by the mean value theorem for integrals we have

$$\left| \int_{(m-1)/n}^{m/n} [F(x) - F(m/n)]dx \right| \leq \frac{1}{n} \frac{L}{n} .$$

If we take the sum over the N points am/n in the interval $b \leq x < c$ we have

$$(2.5.26) \qquad \left| \int_b^c F(x)dx - \frac{1}{n} \sum F(\tfrac{m}{n}) \right| \leq \frac{L}{n} .$$

The function $S(x)$ is of total variation A in (0,1) and hence is the difference between two functions, $S_1(x)$ and $S_2(x)$, monotonic non-decreasing in (0,1), the sum of whose total variation is A. As above we have

$$\left| \int_{(m-1)/n}^{m/n} [S_1(x) - S_1(\tfrac{m}{n})]dx \right| \leq \frac{1}{n} [S_1(\tfrac{m}{n}) - S_1(\tfrac{m-1}{n})]$$

$$\left| \int_b^c S_1(x)dx - \frac{1}{n} \sum S_1(\tfrac{m}{n}) \right| \leq \frac{1}{n} [S_1(1) - S_1(0)] .$$

Since we have a similar inequality for $S_2(x)$ and since both the integral and summation are additive with respect to the function we have

$$(2.5.27) \qquad \left| \int_b^c S(x)dx - \frac{1}{n} \sum S(\tfrac{m}{n}) \right| \leq A/n.$$

We now have merely to recall that $F(x) = f(x) - S(x)$ and to combine inequalities (2.5.26) and (2.5.27) to complete the proof of Lemma 2.5.5.

Hence if C is a contour we see that equations (2.5.16) and (2.5.17) yield by virtue of Lemma 2.5.5

$$(2.5.28) \qquad |g - \frac{1}{n} \log |\tau_n|| \leq M/n, \quad |ng - \log |\tau_n|| \leq M$$

for $|w| = 1$ or z on C, since the integrand in (2.5.16)

satisfies on each γ_j a Lipschitz condition with respect to u or w', uniformly on each closed arc interior to a γ_k. By Lemma 2.5.5 we see that M can be chosen as independent of both n and w, where w lies on an arbitrary closed arc interior to a γ_j, or in other words where z lies on any closed proper sub-arc of any C_j. Consequently, by use of a new correspondence between γ and C, involving new points of C_j as images of the end-points of the γ_j, it follows that these inequalities are valid for suitable choice of M, uniformly for all z on C_j.

When such a new correspondence is set up between γ and C, care must be taken not to alter the geometric significance of the points ζ_m on C. This geometric significance is to be fixed once for all by the original correspondence. For definiteness we choose the points $w_m = \phi(\zeta_m)$ as the n-th roots of unity. The new correspondence, say between γ_1 and C_1, is best broken up into a correspondence (suggested by the original correspondence between γ_1 and C_1) between two arcs of C_1 and corresponding arcs of γ; for w on each such arc of γ the conclusion of Lemma 2.5.5 as just interpreted is valid; the Lipschitz condition with respect to w' on the integrand of (2.5.16) holds uniformly for w on each of these two arcs of γ except for w in the neighborhoods of one end-point of each arc; these two neighborhoods correspond to one-sided neighborhoods of a single point on C_1, a point different from the exceptional point of C_1 under the original correspondence between γ_1 and C_1. Thus the inequalities involving τ_n are established uniformly for all w on γ_1 or γ_j, thus for all z on C_1 or C_j.

By taking exponentials in (2.5.28) we now obtain

$$e^{-M} \leq \frac{e^{ng}}{|\tau_n|} \leq e^{M}, \quad e^{-M} \leq \left|\frac{e^{ng}(w^n-1)}{\omega_{n-1}(z)}\right| \leq e^{M};$$

another form of these inequalities is slightly more convenient

$$|\omega_{n-1}(z)| \leq e^{M+ng}|w^n-1| \leq M_1 e^{ng},$$

(2.5.29)

$$|\omega_{n-1}(z)| \geqq M_2 e^{ng} |w^n - 1|,$$

where $|w| = 1$ and z lies on C. We have proved

LEMMA 2.5.6. If C is a contour, inequalities (2.5.29) are uniformly valid, where $|w| = 1$ and z lies on C.

We now study $\omega_{n-1}(z)$ for z on C_ρ. Equation (2.5.2) may be written

$$\int_C \log |z - \zeta| \, du = \log \rho + g,$$

whence from Lemma 2.5.5

$$\left| \log \rho + g - \frac{1}{n} \log |\omega_{n-1}(z)| \right| \leqq M'/n,$$

(2.5.30) $0 < M_1' e^{ng} \rho^n \leqq |\omega_{n-1}(z)| \leqq M_2' e^{ng} \rho^n$, z on C_ρ.

The derivation of (2.5.30) requires much less than that C be a contour, merely that the function $\log |z - \zeta| = \log |z - \psi(w)|$ satisfy uniformly for all z on C_ρ a Lipschitz condition with respect to u or w, on each γ_j. It is sufficient (compare Lemma 2.5.9 below) if $|z - \psi(w)|$ satisfies such a Lipschitz condition, for which it is sufficient if $\psi(w)$ itself satisfies such a condition; the existence and continuity of $\psi'(w)$ on $|w| = 1$ are ample to ensure this; compare inequality 2.5.37 below.

Merely a part of the conclusions in (2.5.29) and (2.5.30) is summarized in

THEOREM 2.5.7. Let C be a contour, and let the points z_m be equally distributed on C. Then we have

$$\left| \frac{\omega_n(z)}{\omega_n(t)} \right| \leqq M_0/\rho^n, \quad z \text{ on } C, \ t \text{ on } C_\rho,$$

where M_0 is a constant independent of n, z, and t.

Theorem 2.5.7 is fundamental in our study of degree of approximation, in both Problem α and Problem β ; in fact this theorem in combination with the Lagrange-Hermite interpolation formula is our principal tool in Problem β . The concept of contour is so important in our study of Problem β that we go into further detail concerning its properties. In this connection we have

THEOREM 2.5.8. Let the function $|[\psi(w) - \psi(w')]/(w-w')|$ be uniformly bounded from zero for $|w| = 1$ and $|w'| = 1$, and on each closed arc of γ interior to a γ_j let $\psi''(w)$ be continuous. Then C is a contour.

In this theorem and also in Theorem 2.5.10 below we naturally assume an interpretation similar to that of Definition 2.5.2 relative to the various choices of the maps of C on γ.

In the proof of Theorem 2.5.8 it is convenient to have for reference

LEMMA 2.5.9. If on a closed finite interval of the axis of reals or a closed arc of the unit circle the function $f(x_1, x_2)$ is positive and satisfies a Lipschitz condition of given order α, $0 < \alpha \leq 1$, in x_1 uniformly with respect to x_2:

$$|f(x_1, x_2) - f(x_1', x_2)| \leq L\,|x_1 - x_1'|^{\alpha},$$

then $\log f(x_1, x_2)$ satisfies on that interval or arc a Lipschitz condition of order α in x_1, uniformly with respect to x_2.

Since $f(x_1, x_2)$ is positive and continuous on the

given point set, that function is uniformly bounded from zero there: $f(x, x_2) > m > 0$. We have the expansion

$$\log f(x_1, x_2) - \log f(x_1', x_2)$$

$$(2.5.31) \qquad = \frac{f(x_1,x_2) - f(x_1', x_2)}{f(x_1', x_2)}$$

$$\left[1 - \frac{1}{2} \left[\frac{f(x_1,x_2) - f(x_1', x_2)}{f(x_1', x_2)} \right] + \dots \right],$$

which is valid provided $|[f(x_1,x_2) - f(x_1', x_2)]/f(x_1',x_2)| < 1$; this inequality is satisfied uniformly for x_1' sufficiently near x_1. Furthermore, it follows from the boundedness of $f(x_1,x_2)$ that the given Lipschitz condition on $\log f(x_1,x_2)$ need be established only for $|x_1'-x_1|$ sufficiently small by the same argument which we used in connection with (2.5.25). For $|x_1'-x_1| < \delta$ we have from (2.5.31) and from the given Lipschitz condition on $f(x_1,x_2)$

$$|\log f(x_1,x_2) - \log f(x_1',x_2)| \leq$$

$$\frac{M}{m} |x-x_1'|^{\alpha} \left[1 + \frac{M}{2m} | x-x_1'|^{\alpha} + \dots \right],$$

provided $\delta^{\alpha} < m/M$. We read off at once the given Lipschitz condition on $\log f(x_1,x_2)$, including uniformity with respect to x_2.

In proving Theorem 2.5.8 we shall use the mean value theorem of Darboux [1876] in a later and generalized form due to Walsh and Sewell [1940]:

> Let $f(z)$ and $f'(z)$ be continuous along the line segment from z_0 to z_1; then we have
> $$(2.5.32) \quad f(z_1)-f(z_0) = \lambda(z_1-z_0)\cdot f'[z_0+ \epsilon(z_1-z_0)],$$
> where $0 \leq \epsilon \leq 1$, $|\lambda| \leq 1$.

If $f''(z)$ is also continuous, replacement of $f(z)$ by $f(z)$-

$(z-z_1)f'(z)$ yields

$$(2.5.33)\ f(z_1) = f(z_0) + f'(z_0)(z_1-z_0) + \lambda(z_1-z_0)^2$$
$$f''[z_0 + \epsilon(z_1-z_0)].$$

By integration along the segment z_0 to z_1 we may write

$$f(z_1) - f(z_0) = \int_{z_0}^{z_1} f'(z)\, dz,$$

$$(2.5.34)\ |f(z_1) - f(z_0)| \leqq |z_1 - z_0| \cdot \max |f'(z)|\ ;$$

this is essentially equation (2.5.32). If the curve (assumed to have a continuously turning tangent) on which we study f(z) is not a line segment inequality (2.5.34) remains valid provided we replace $|z_1 - z_0|$ by the length of arc, whence

$$(2.5.35)\ f(z_1) - f(z_0) = \lambda f'(z') \int_{z_0}^{z_1} |dz|,$$

where z' is a point on the arc. Likewise equation (2.5.33) may be replaced by

$$(2.5.36)\ f(z_1) = f(z_0) + f'(z_0)(z_1-z_0) + \lambda f''(z') \left[\int_{z_0}^{z_1} |dz|\right]^2,$$

where $|\lambda| \leqq 1$, where the integral is taken over the arc considered, and where z' lies on this arc. If the function f(z) is studied on a <u>circular arc</u> on which z_0 and z_1 lie we may use the inequality

$$\int_{z_0}^{z_1} |dz| \leqq \frac{\pi}{2} |z_1 - z_0|;$$

we assume here that the arc considered is not greater than a semicircle; in the contrary case we construct the subsequent proof in several steps as in treating (2.5.25). Thus (2.5.35) takes the form

$$(2.5.38)\ f(z_1) = f(z_0) + f'(z_0)(z_1-z_0) + \lambda\left(\frac{\pi}{2}\right)^2 (z_1 - z_0)^2 f''(z'),$$

with $|\lambda| \leq 1$.

Under the hypothesis of Theorem 2.5.8 we now write

$$(2.5.39) \qquad \Psi(w,w') = \frac{\psi(w) - \psi(w')}{w - w'}$$

$$(2.5.40) \qquad \Psi'(w,w') = \frac{(w-w')\psi'(w) - [\psi(w)-\psi(w')]}{(w - w')^2}$$

where the accent indicates differentiation with respect to w. Let the points w and w' both lie on a closed sub-arc interior to γ_j; then it follows from (2.5.38) and (2.5.40) that $\Psi(w,w')$ is uniformly bounded unless $w = w'$. The function $\Psi(w,w')$, considered as a function of w, is continuous even in the point $w = w'$ provided we set $\Psi(w,w') = \psi'(w')$.

In the derivation of (2.5.35) and (2.5.36), the existence or non-existence of f'(z) or f"(z) at a single point may be disregarded, provided f'(z) or f"(z) exists elsewhere, is uniformly bounded, and is continuous. It then follows from 2.5.37 that $\Psi(w,w')$ satisfies a Lipschitz condition

$$|\Psi(w_1,w') - \Psi(w_2,w')| \leq L|w_1 - w_2|$$

for w_1, w_2, w' on the closed sub-arc interior to γ_j, or even if w' lies on a closed sub-arc interior to a γ_k, $k \neq j$. The Lipschitz condition in w on the function (2.5.18) uniformly with respect to w' now follows from Lemma 2.5.9 by virtue of the inequality

$$\Big| |\Psi(w_1,w')| - |\Psi(w_2,w')| \Big| \leq |\Psi(w_1,w') - \Psi(w_2,w')|;$$

the fact that $\Psi(w,w')$ is bounded from zero is part of our hypothesis.

The existence of $\psi''(w)$ on $|w| = 1$ implies the continuity of $\psi'(w)$ there, hence implies that C_j has a continuously turning tangent. It follows from the method of proof of (2.5.21) that $\psi'(w)$ satisfies a Lipschitz condition on the arc considered. It remains, however, to prove the boundedness of the function (2.5.18) for $|w| = 1$ and $|w'| \geq 1$. We proceed as in the proof and application of Lemma 2.5.4; the inequality $\psi'(w) \neq 0$ is a consequence of the hypothesis that $\Psi(w,w')$ is bounded from zero. It now follows from the theorem of Kellogg [1912] that $\psi'(w)$ exists and is continuous in the two-dimensional sense in the neighborhood of $|w| = 1$. Since $\psi'(w)$ is different from zero on $|w| = 1$, it follows that $\psi'(w)$ is different from zero in the neighborhood of $|w| = 1$. Use the transformations $w_1 = 1/w$, $w_1' = 1/w'$ and then equation (2.5.32) for the segment $w_1 w_1'$ shows that the function

$$\frac{\psi(w) - \psi(w')}{w - w'}$$

is uniformly bounded for $|w| = 1$ and $|w'| = 1$ in the neighborhood of points of continuity of $\psi(w)$; we use here the boundedness of $\psi'(w)$ and the fact that in the derivation of (2.5.32) the boundedness and continuity of the derivative are sufficient even if the derivative fails to exist at a single point. Further use of the transformations $w_1 = 1/w$, $w_1' = 1/w'$ provides us with a convenient path, namely, the line segment joining w_1 and w_1'; the transformations $z = \psi(w)$, $z' = \psi(w')$ then show by means of (2.4.35) that the function

$$\frac{w - w'}{\psi(w) - \psi(w')} = \frac{\phi(z) - \phi(z')}{z - z'}$$

is uniformly bounded at points of continuity of $\psi(w)$, for $|w| = 1$ and $|w'| \geq 1$. Thus the function (2.5.18) is locally bounded for $|w| = 1$ and $|w'| \geq 1$ in the neighborhood of points of continuity of $\psi(w)$. Theorem 2.5.8 is established.

Still another sufficient condition may be mentioned:

THEOREM 2.5.10. If the function

$$(2.5.41) \qquad \log \left[\frac{\psi(w) - \psi(w')}{w - w'} \right]$$

satisfies on each closed arc interior to ever γ_j a Lipschitz condition, then C is a contour.

The Lipschitz condition on the function (2.5.41) implies at once the corresponding Lipschitz condition on the function (2.5.18), the real part of (2.5.41). By the method of proof of (2.5.21) and by Lemma 2.5.3 it follows that $\psi'(w)$ exists on any closed arc interior to γ_j and satisfies a Lipschitz condition there, and moreover that $\psi'(w)$ is different from zero. The boundedness of the function (2.5.18) follows from the reasoning given in the proof of Lemma 2.5.4. The proof of Theorem 2.5.10 is complete.

The exposition of this section follows closely that of Walsh and Sewell [1940]. Inequalities (2.5.29) are due to Curtiss [1935] for C a single Jordan curve whose mapping function $\psi(w)$ has a second derivative satisfying a Lipschitz condition.

§2.6. EXERCISES. 2.6.1. If $|P_n(z)| \leq 1$, $|z| \leq 1$, then $|P_n'(z)| \leq n$, $|z| \leq 1$. This bound is obtained only by $P_n(z) = \alpha z^n$, $|\alpha| = 1$. [Bernstein, 1926, p. 45]

SUGGESTION. Use the fact that if

$$T_n(\theta) = a_0 + \sum_{m=1}^{n} (a_m \cos m\theta + b_m \sin m\theta)$$

is an arbitrary trigonometric sum of order n, then $|T_n(\theta)| \leq 1$ for all θ implies $|T_n'(\theta)| \leq n$ for all θ. [Bernstein, 1926, p. 39]

2.6.2. If $|P_n(z)| \leq M$ on the circle $|z-a| \leq b$, then $|P_n'(z)| \leq Mn/b$, $|z-a| \leq b$.

2.6.3. If $|P_n(z)| \leq M$ on the segment $-1 \leq z \leq 1$, then $|P_n'(z)(1-z^2)^{1/2}| \leq Mn$, $-1 \leq z \leq 1$. [Bernstein, 1926, p. 38]

2.6.4. If $|P_n(z)| \leq M$ for z on an arbitrary line

segment of length 2 l, then $|P_n'(z)| \leqq Mn^2 / l$, for z on this segment. This bound cannot be improved.

SUGGESTION. Use the fact that $|P_n(z)| \leqq 1$, $-1 \leqq z \leqq 1$, implies $|P_n'(z)| \leqq n^2$, $-1 \leqq z \leqq 1$. The bound is attained only by $P_n(z) = \alpha \cos (n \cos^{-1} z)$, $|\alpha| = 1$. [Markoff, 1889]

2.6.5. Let C be an ellipse with semi-axes of lengths a,b,a $\geqq$ b. Then $|P_n(z)| \leqq M$, z on C, implies $|P_n'(z)| \leqq Mn/b$, z on C. [Sewell, 1937d]

2.6.6. Let C be a cardioid whose polar equation is $r = a(1 - \cos \theta)$. Then $|P_n(z)| \leqq M$, z on C, implies $|P_n'(z) \cdot \sin \theta/2| \leqq Mn/a$, z on C. [Sewell, 1938]

2.6.7. Let C be a rose curve whose polar equation is $r = a \sin m\theta$. Then $|P_n(z)| \leqq M$, z on C, implies $|P_n'(z)| \leqq M(m+1)n/a$, z on C. [Sewell, 1938]

2.6.8. Let C be an epicycloid or a hypocycloid whose parametric equations are

$$x = (a \pm b) \cdot \cos \phi \mp b \cdot \cos \frac{a \pm b}{b} \phi ,$$

$$y = (a \pm b) \cdot \sin \phi - b \cdot \sin \frac{a \pm b}{b} \phi ,$$

where the upper signs belong to the epicycloid and the lower signs to the hypocycloid, and b is a factor of a. Then $|P_n(z)| \leqq M$, z on C, implies $|P_n'(z) \sin (a\phi/2b)| \leqq Mn/(2b)$, z on C. [Sewell, 1938]

2.6.9. Let C be the limaçon whose polar equation is $r = a + b \cdot \cos \theta$. Then $|P_n(z)| \leqq M$, z on C, implies $|P_n'(z)| \leqq M \cdot 2n / | |a| - |b| |$, z on C. [Sewell, 1939b]

2.6.10. Let C be the lemniscate whose polar equation is $r^2 = a^2 \cos 2\theta$. Then $|P_n(z^2)| \leqq M$, z on C, implies $|P_n'(z^2)| \leqq M \cdot 4n \cdot |(\cos 2\theta)^{1/2}| / a$, z on C. [Sewell, 1939b]

2.6.11. In the real (x,y)-plane let C be the ellipse

$$\frac{x^2}{a^2} + \frac{y^2}{b^2} = 1.$$

Let $P(x,y)$ be a polynomial of degree n in the variables x and y together. Then $|P(x,y)| \leq M$ at all points of C implies

$$\left| \frac{d\,P(x,y)}{ds} \right| \leq K\,M\,n$$

at all points of C, where K is an absolute constant, and s is the arc length along C. [Jackson, 1937]

2.6.12. Let C be the curve $x^4 + y^4 = 1$. Then under the hypothesis of Ex. 2.4.11 we have the same conclusion. [Jackson, 1937]

2.6.13. Let C be a Jordan curve such that through each point a of C can be drawn a circle $\gamma(a)$ of radius $\delta \geq \epsilon > 0$, with ϵ fixed for all points a; suppose the closure of $\gamma(a)$ lies in $\overline{C}$. Then $|P_n(z)| \leq M$, z on C, implies $|P_n'(z)| \leq KMn$, z on C, where K is a constant depending only on C and ϵ. [Jackson, 1930a]

2.6.14. Let C be a Jordan curve such that from each point a of C can be drawn a closed line segment of length $\delta \geq \epsilon > 0$, which lies in $\overline{C}$, where ϵ is independent of a. Then $|P_n(z)| \leq M$, z on C, implies $|P_n'(z)| \leq KMn^2$, z on C, where K is a constant depending only on C and ϵ. [Jackson, 1931]

2.6.15. Let C consist of the two circles

$$|z + 1| = 1, \qquad |z - 1| = 1$$

Then $|P_n(z)| \leq M$, z on C, implies $|P_n'(0)| \leq KM \log n$, z on C, where K is an absolute constant. [Szegö, 1925]

SUGGESTION. Use the fact that

$$w = \psi(z) = i\,\frac{1 + e^{-\frac{i\pi}{z}}}{1 - e^{\frac{i\pi}{z}}}$$

maps the exterior of C on $|w| > 1$ so that the points at infinity in the two planes correspond to each other.

2.6.16. Let C be a contour and let the function $\log|z - \zeta| = \log|z - \psi(w)|$ satisfy a Lipschitz condition in u or w on each γ_j, uniformly for all z on C_ρ. Then inequalities (2.5.28) are valid. [Walsh and Sewell, 1940]

2.6.17. Let $\psi''(w)$ be integrable and bounded, then C is a contour. [Walsh and Sewell, 1940]

SUGGESTION. Compare proof of Theorem 2.6.8.

2.6.18. Let C be an analytic Jordan arc; let the function $\psi(z)$ be analytic on a one-sided neighborhood of C and continuous in the corresponding closed region R. For a particular interior point z' of C let $\psi'(z')$ exist on C in the one-dimensional sense, and suppose

$$\lim_{z \to z',\ z \text{ in } R} \psi'(z)$$

is bounded. Then the difference quotient $[\psi(z)-\psi(z')]/(z-z')$ is bounded in the two-dimensional sense for z interior to R, and the derivative $\psi'(z')$ is a two-dimensional derivative. [Walsh and Sewell, 1940]

2.6.19. Let C be a Jordan curve composed of a finite number of analytic arcs meeting in corners of exterior openings $\mu_j \pi$, $1 \leq \mu_j < 2$, and let

$$\iint_{\overline{C}} |P_n(z)|^p \, dS \leq L^p, \quad p > 0,$$

where $P_n(z)$ is an arbitrary polynomial of degree n in z. Then

$$|P_n(z)| \leq KL\, n^{2t/p}, \quad z \text{ on } \overline{C},$$

where K is a constant depending on C and p but independent of $P_n(z)$, n, and z, and t is the type of C. The exponent of n cannot be improved for t = 1 and p = 2.

§2.7. OPEN PROBLEMS. This chapter presents an abundance of interesting open problems. In Theorem 2.1.4 the "best constant" K(C) is worthy of investigation for its own sake. As far as our applications here are concerned it is sufficient to know that K(C) depends only on C, however its exact value for various types of curves and sets is of interest. It is clear from Exs. 2.6.1,

2.6.5, and 2.6.4 that the actual value of K(C) is known for the circle, the ellipse, and the line segment; still there is ample opportunity for further investigation. Of course an upper bound is easily obtained by an analysis of the proof but the "best value" and the geometric significance of this constant require much more delicate tools.

Obviously we have a similar problem concerning the constant of Theorem 2.2.1. It is to be observed that Ex. 2.6.19 yields an inequality for surface integrals similar to that of Theorem 2.2.1 for line integrals, however the result does not include the case $t = 2$. The extension of Ex. 2.6.19 to further sets, for instance to curves with cusps, is an interesting open problem.

In both Theorems 2.1.4 and 2.2.1 there is also the problem of determining the "best" exponents of n. Of course our examples show that in certain cases the exponents are actually the best possible, but there are many important cases yet to be studied.*

The investigation of the behavior of equally distributed points and $\omega_{n-1}(z)$ for curves other than contours is beyond the scope of this treatise; however, results similar to those of §2.5 for more general configurations have immediate application to the theory of approximation. Furthermore in connection with a contour a study of necessary conditions is highly desirable, as well as a more thorough investigation of sufficient conditions.

There is also the problem of the behavior of the equally distributed points and $\omega_{n-1}(z)$ on (i) a single Jordan curve which does not have the continuity properties of a contour, and on (ii) configurations consisting of several Jordan arcs and curves which may or may not

* See Ahlfors [1930]; Caratheodory [1932]; Keldysch and Lavrentieff [1937]; Phragmen and Lindelöf [1908]; Visser [1932]; Warschawski [1932, 1932a, 1934].

have points in common. A special case of (i) is a Jordan curve of which the function defining the parametric representation does not have an integrable and bounded second derivative [compare Kellogg, 1912; Seidel, 1931]. In Theorem 2.5.8 the requirement of the continuity of $\psi''(w)$ can be somewhat lessened [Walsh and Sewell, 1940], in fact it is sufficient if $\psi''(w)$ is integrable and bounded, by obvious modifications of the proof. We mention also the special case of a curve with corners such as a square or a triangle. In connection with (ii) various special configurations are of interest in themselves as well as being indicative of the situation for the more general problem; among these special configurations we mention (a) two equal circles which are tangent externally or which intersect in two points, (b) two equal segments of the same line, (c) two equal segments which form opposite sides of a rectangle, (d) two equal segments which bisect each other perpendicularly.

Chapter III

TCHEBYCHEFF APPROXIMATION

§3.1. JACKSON'S THEOREM ON TRIGONOMETRIC APPROXIMATION. The theorems of Chapter II form the foundation for our study of degree of approximation by polynomials; however our proof of the direct theorems of Problem α depends upon an important result on degree of approximation by trigonometric sums. This result belongs to the theory of functions of a real variable but, as mentioned in the preface, approximation by trigonometric sums and approximation by polynomials in z on the unit circle are closely related, in fact so closely that it seems advisable to set forth the details of the proof of this important theorem. In the present section we are occupied with developing a tool to be applied to complex approximation whereas in §8.3 in connection with Problem β our purpose is to exhibit the reciprocal relation between trigonometric approximation and approximation by polynomials.

By a <u>trigonometric sum of the n-th order</u> is meant an expression of the form

$$\begin{aligned} S_n(\theta) &= a_0 + a_1 \cos \theta + a_2 \cos 2\theta + \dots + a_n \cos n\theta \\ &\quad + b_1 \sin \theta + b_2 \sin 2\theta + \dots + b_n \sin n\theta \\ &= \sum_{m=0}^{n} (a_m \cos m\theta + b_m \sin m\theta), \end{aligned}$$

where the a_m and b_m are constants. Let $f(\theta)$ be periodic of period 2π and integrable together with its square; if

$$a_n = \frac{1}{2\pi} \int_{-\pi}^{\pi} f(\theta) \cos n\theta \, d\theta, \quad b_n = \frac{1}{2\pi} \int_{-\pi}^{\pi} f(\theta) \sin n\theta \, d\theta ,$$

then $S_n(\theta)$ is the sum of the first $2n + 1$ terms of the Fourier development of the function $f(\theta)$. If $f(\theta)$ satisfies a Lipschitz condition of some positive order this development converges to the function and hence the partial sums approximate the function. However we obtain a better degree of approximation by considering not the partial sums themselves but a particular summation of these partial sums. The theorem is

THEOREM 3.1.1. Let $f(\theta)$ be periodic of period 2π, and let $f(\theta)$ belong to the class $L(k-1, \alpha)$ on $-\infty < \theta < \infty$ that is to say let $f(\theta)$ have a $(k-1)$-th derivative satisfying a Lipschitz condition of given order α, $0 < \alpha \leq 1$,

$$(3.1.1) \quad |f^{(k-1)}(\theta_1) - f^{(k-1)}(\theta_2)| \leq L\,|\theta_1 - \theta_2|^\alpha ,$$

for all real θ. Then for each n, $n = 1, 2, \ldots$, there exists a trigonometric sum $S_n(\theta)$ of order n, such that

$$(3.1.2) \quad |f(\theta) - S_n(\theta)| \leq ML/n^{k-1+\alpha} ,$$

where M is a constant depending only on k.

Although the statement is in the form of an existence theorem the method of proof establishes much more. In fact we study the form of the sum $S_n(\theta)$, which form is of significance in §3.2 when we apply Theorem 3.1.1 to our study of degree of approximation in the complex domain. The proof consists in forming a particular trigonometric sum and proving the inequality (3.1.2). We shall write down an integral expression and show that it is not only a trigonometric sum of satisfactory order but that it is also a summation of the Fourier development of $f(\theta)$.

We define an approximating function $I_m(\theta)$ as follows:

$$I_m(\theta) = h_m \int_{-\pi/2}^{\pi/2} [(-1)^{k+1} f(\theta+2ku) + (-1)^k k f(\theta+2(k-1)u) + \ldots + kf(\theta+2u)] [\frac{\sin mu}{m \sin u}]^{2p} du,$$

where the numerical coefficients in the integrand are the binomial coefficients corresponding to the exponent k, the last being omitted, where p is the smallest integer for which $2p - k - \alpha > 0$, m is the largest integer for which $k(m-1)$ does not exceed n, with n preassigned, and where

$$\frac{1}{h_m} = \int_{-\pi/2}^{\pi/2} [\frac{\sin mu}{m \sin u}]^{2p} du.$$

Since

$$f(\theta) = h_m \int_{-\pi/2}^{\pi/2} f(\theta) [\frac{\sin m u}{m \sin u}]^{2p} du,$$

the difference $I_m(\theta) - f(\theta)$ may be written in a form which differs from that of $I_m(\theta)$ only in that it has one more term, $-f(\theta)$, in the first factor of the integrand. The resulting factor is so constituted that when it is developed by Taylor's theorem to terms of the k-th degree in u all the terms of degree lower than the p-th in the expressions combine so that their sum is identically zero. Thus we can apply the inequality (3.1.1); in fact it follows that the whole expression does not exceed a constant multiple of $L|u|^{\alpha} u^{k-1}$ in absolute value. This, of course, explains the form of the first part of the integrand in the definition of $I_m(\theta)$. The fact that the integrand reduces as described above is easily verified by writing out the expansion at length; we then have merely to show that the function

$$S_{k,i} = t^i - k(t+1)^i + \ldots + (-1)^{k-1}(t+k-1)^i + (-1)^k(t+k)^i,$$

where the numerical coefficients are still the binomial coefficients for the exponent k, vanishes identically whenever $0 \leq i \leq k-1$, i being an integer; what is needed is only the fact that it vanishes when $t = 0$. Since

$$\frac{dS_{k,i}(t)}{dt} \equiv i\, S_{k,i-1}(t),$$

$S_{k,i-1}(t)$ vanishes identically if $S_{k,i}(t)$ does; it is sufficient to prove for each value of k that $S_{k,k-1}(t)$ is identically zero. We proceed by induction. Suppose $S_{k-1,k-2}(t) \equiv 0$ for a given value of k; it is certainly true for $k = 2$. Then

$$\frac{d}{dt} S_{k-1,k-1}(t) \equiv (k-1)\, S_{k-1,k-2}(t) \equiv 0,$$

hence $S_{k,k-1}(t)$, which is equal to $S_{k-1,k-1}(t) - S_{k-1,k-1}(t+1)$, is identically zero. The induction is complete.

Thus it follows that

$$|I_m(\theta) - f(\theta)| \leq \frac{KL \int_0^{\pi/2} u^{k-1+\alpha} \left[\frac{\sin mu}{m \sin u}\right]^{2p} du}{\int_0^{\pi/2} \left[\frac{\sin mu}{m \sin u}\right]^{2p} du},$$

where K depends only on k. By using the fact that $0 < \sin u < u$ for $0 < u \leq \pi/2$ and making the substitution $mu = t$, we have

$$\int_0^{\pi/2} \left[\frac{\sin mu}{m \sin u}\right]^{2p} du > \int_0^{\pi/2} \left[\frac{\sin mu}{mu}\right]^{2p} du$$

$$= \frac{1}{m} \int_0^{m\pi/2} \frac{\sin^{2p} t}{t^2}\, dt$$

$$\geq \frac{1}{m} \int_0^{\pi/2} \frac{\sin^{2p} t}{t^2}\, dt.$$

On the other hand, $(\sin u)/u$ decreases monotonically as u varies from 0 to $\pi/2$, so that

$$\frac{\sin u}{u} > \frac{\sin \pi/2}{\pi/2} = \frac{2}{\pi}, \quad \frac{1}{\sin u} < \frac{\pi}{2} \cdot \frac{1}{u}$$

throughout the interior of this interval. Hence

$$\int_0^{\pi/2} u^{k-1+\alpha}\left[\frac{\sin mu}{m \sin u}\right]^{2p} du \leqq \left(\frac{\pi}{2}\right)^{2p}\int_0^{\pi/2} u^{k-1+\alpha}\left[\frac{\sin m u}{m u}\right]^{2p} du$$

$$= \left(\frac{\pi}{2}\right)^{2p}\int_0^{m\pi/2} \frac{t^{k-1+\alpha}}{m^{k-1+\alpha}} \frac{\sin^{2p} t}{t^{2p}} \frac{dt}{m}$$

$$\leqq \frac{1}{m^{k+\alpha}}\left(\frac{\pi}{2}\right)^{2p}\int_0^{\infty} \frac{\sin^{2p} t}{t^{2p-k+1-\alpha}} dt,$$

and this last integral exists since $2p-k-\alpha > 0$.

Now let us examine the function $I_m(\theta)$. We shall show that

$$(3.1.3)\quad I_m(\theta) = \int_{-\pi}^{\pi} f(t)\, T_n(t-\theta)\, dt,$$

where

$$(3.1.4)\quad T_n(t-\theta) = \sum_{\nu=0}^{n} A_{n,\nu} \cos \nu(t-\theta),$$

the numbers $A_{n,\nu}$ being constants independent of $f(\theta)$. It is sufficient to show that each of the k terms of which $I_m(\theta)$ is composed has this form. Consider any one of these terms, say that containing $f(\theta+2ru)$; of course the binomial coefficient and the factor $\pm h_m$ may be left out of account. The change of variable $t = \theta + 2ru$ gives

$$\int_{-\pi/2}^{\pi/2} f(\theta+2ru)\left[\frac{\sin mu}{m \sin u}\right]^{2p} du = \frac{1}{2r}\int_{x-r\pi}^{x+r\pi} f(t)\left[\frac{\sin m\frac{t-\theta}{2r}}{m \sin\frac{t-\theta}{2r}}\right]^{2p} dt.$$

Because of the periodicity of the functions involved, this is, except for the irrelevant constant factor $1/(2r)$, the same as

$$\int_{-r\pi}^{r\pi} f(t)\left[\frac{\sin m\,\frac{t-\theta}{2r}}{m \sin \frac{t-\theta}{2r}}\right]^{2p} dt.$$

By breaking up the interval of integration into r intervals each of length 2π and reducing these to a common interval by variable changes we can bring this integral into the form

$$(3.1.5) \qquad \int_{-r\pi}^{-(r-2)\pi} f(t) \sum_{j=0}^{r-1} \left[\frac{\sin m(\frac{t-\theta}{2r} + \frac{2j\pi}{2r})}{m \sin (\frac{t-\theta}{2r} + \frac{2j\pi}{2r})} \right]^{2p} dt.$$

It is true that

$$(3.1.6) \qquad \left[\frac{\sin \frac{mv}{2}}{m \sin \frac{v}{2}} \right]^2$$

is a finite trigonometric sum in v of order m-1 involving only cosines. For we have the identities

$$\frac{\sin^2 \frac{mv}{2}}{\sin^2 \frac{v}{2}} = \frac{1 - \cos m v}{1 - \cos v},$$

$$1 - \cos m v = \sum_{j=0}^{m-1} [\cos jv - \cos (j+1) v],$$

$$\cos jv - \cos (j+1)v = (1 - \cos v) - \sum_{k=1}^{j} [\cos(k-1)v - 2 \cos kv + \cos(k+1)v],$$

$$\cos (k-1) v - 2 \cos k v + \cos (k+1)v =$$

$$[\cos (k-1)v + \cos (k+1)v] - 2 \cos k v$$

$$= 2 \cos k v . \cos v - 2 \cos k v$$

$$= 2 \cos k v . (\cos v - 1);$$

thus 1-cos mv is divisible by (1-cos v), the quotient being a trigonometric sum in v of order m-1 involving only cosines. Also because of the identity

$$\cos jv \cos kv = \frac{1}{2} [\cos (j+k)v + \cos (j-k)v],$$

it is clear that the p-th power of (3.1.6) is a trigonometric sum of order p(m-1) involving only cosines.

Thus each term under the sign of summation in the above integral has the form

$$\sum_{i=0}^{p(m-1)} B_i \cos i \left(\frac{t-\theta}{r} + \frac{2j\pi}{r}\right),$$

where the numbers B_i are constants. On performing the summation with regard to j, a double sum is obtained which can be written thus:

$$(3.1.7) \qquad \sum_{i=0}^{p(m-1)} \left[B_i \left[\cos i \frac{t-\theta}{r} \sum_{j=0}^{r-1} \cos \frac{2ij\pi}{r} - \sin i \frac{t-\theta}{r} \sum_{j=0}^{r-1} \sin \frac{2ij\pi}{r} \right] \right]$$

But

$$\sum_{j=0}^{r-1} \sin \frac{2ij\pi}{r} = 0,$$

as we shall show by considerations of symmetry. Of course we have sin 0 = 0. Also since

$$\frac{i\ 2\ (r-j)\pi}{r} = i2\pi - \frac{i2j\pi}{r},$$

each term corresponding to an index j between 0 and r/2 is paired with a numerically equal term of opposite sign. If r is odd, all the terms are thus accounted for; if r is even, the term which remains by itself is

$$\sin \frac{2i(r/2)\pi}{r} = \sin i\pi = 0.$$

Furthermore

$$\sum_{j=0}^{r-1} \cos \frac{2ij\pi}{r}$$

is zero unless i is divisible by r. We have the well known identity

$$\frac{1}{2} + \cos u + \cos 2u + \ldots + \cos nu = \frac{\sin (n + \frac{1}{2}) u}{2 \sin \frac{1}{2} u},$$

where the right-hand member is defined by continuity at points where the denominator vanishes. If $u = i2\pi/r$ then $mu = i2m\pi/r$, and

$$\frac{1}{2} + \cos \frac{i2\pi}{r} + \cos \frac{i4\pi}{r} + \ldots + \cos \frac{i2n\pi}{r} = \frac{\sin[(n+\frac{1}{2})\frac{i2\pi}{r}]}{2 \sin[\frac{1}{2} \frac{i2\pi}{r}]},$$

But by the symmetry pointed out above the terms on the

left are respectively the same as those of the sum

$$\frac{1}{2} + \cos\frac{i2(r-1)\pi}{r} + \cos\frac{i2(r-2)\pi}{r} + \dots + \cos\frac{i2(r-n)\pi}{r}.$$

If $r = 2n + 1$, and if i is not divisible by r, addition of the two sums gives

$$\sum_{j=0}^{r-1}\cos\frac{2ij\pi}{r} = \frac{\sin\left[(n+\frac{1}{2})\frac{i2\pi}{r}\right]}{\sin\left[\frac{1}{2}\frac{i2\pi}{r}\right]},$$

and in the last expression the numerator is zero, since $(n+\frac{1}{2}) = r/2$ so that $(n+\frac{1}{2})(i2\pi/r) = i\pi$, while the denominator is different from zero. If $r = 2n$, the term $\cos(i2n\pi/r)$ occurs in each sum; combination of the two sums, with subtraction of the redundant term, yields

$$\sum_{j=0}^{r-1}\cos\frac{2ij\pi}{r} = \frac{\sin\left[(n+\frac{1}{2})\frac{i2\pi}{r}\right]}{\sin\left[\frac{1}{2}\frac{i2\pi}{r}\right]} - \cos\frac{i2n\pi}{r}$$

$$= \frac{\sin\frac{i2n\pi}{r}\cos\frac{i\pi}{r} + \cos\frac{i2n\pi}{r}\sin\frac{i\pi}{r}}{\sin\left[\frac{i\pi}{r}\right]} - \cos\frac{i2n\pi}{r}$$

$$= \sin\frac{i2n\pi}{r}\cot\frac{i\pi}{r} = \sin i\pi \cot\frac{i\pi}{r} = 0,$$

the assumption being still that i is not divisible by r.

Thus the expression (3.1.7) contains no sines of multiples of $(t-\theta)/r$ at all and cosines of only such multiples of $(t-\theta)/r$ as are at the same time integral multiples of $t-\theta$. It has the same form as the right-hand side of (3.1.4), with the coefficients for which the second subscript is greater than n/r all equal to zero. As the integrand in (3.1.5), regarded as a function of t is now seen to have the period 2π, the interval of integration may equally well be taken as that from $-\pi$ to π. Now to establish (3.1.3) and (3.1.4) it remains only to combine the terms corresponding to the several values of r. The proof of Theorem 3.1.1 is complete.

In order to study the coefficients A_n, let us set $f(\theta) = \cos\nu\theta$. Then

$$\left| \frac{d^k}{d\theta^k} \cos \nu\theta_2 - \frac{d^k}{d\theta^k} \cos \nu\theta_1 \right| \leqq \nu^{k+1} |\theta_2 - \theta_1|,$$

as is seen at once by applying the mean-value theorem to the k-th derivative. Therefore

$$\left| \cos \nu\theta - \int_{-\pi}^{\pi} T_n(t-\theta).\cos \nu t \,.\, dt \right| \leqq \frac{g\,\nu^{k-1}}{n^{k-1}},$$

where g is a constant. On the other hand it follows from (3.1.4) that for $\nu \geqq 1$

$$\int_{-\pi}^{\pi} T_n(t-\theta)\cdot\cos \nu t\cdot dt = \pi A_{n,\nu} \cos \nu\theta,$$

so that

$$|\cos \nu\theta - \pi A_{n,\nu} \cos \nu\theta| \leqq \frac{g\,\nu^{k+1}}{n^{k+1}}$$

for all values of θ, or, setting $\theta = 0$, in particular,

$$(3.1.8) \qquad |1 - \pi A_{n,\nu}| \leqq \frac{g\,\nu^{k+1}}{n^{k+1}} .$$

The same reasoning applied when $\nu = 0$ yields $A_{n,0} = 1/2\pi$.

It appears on substituting (3.1.4) in (3.1.3) that for any function $f(\theta)$ the trigonometric sum $S_n(\theta)$ of the theorem has the form

$$\begin{aligned} S_n(\theta) &= \frac{1}{2\pi} \int_{-\pi}^{\pi} f(t)dt \\ &\quad + \sum_{\nu=1}^{n} \{A_{n,\nu} [\cos \nu\theta \int_{-\pi}^{\pi} f(t)\cos \nu t\, dt + \\ &\qquad\qquad \sin \nu\theta \int_{-\pi}^{\pi} f(t) \sin \nu t\, dt]\} \\ &= a_0 + \sum_{\nu=1}^{n} d_{n,\nu} (a_\nu \cos \nu\theta + b_\nu \sin \nu\theta), \end{aligned}$$

where a_ν and b_ν are the Fourier coefficients of $f(\theta)$, and the numbers $d_{n,\nu} = \pi A_{n,\nu}$ are independent of $f(\theta)$. It is clear that the numbers $d_{n,\nu}$ depend on k; we say that they are of order k in this case. Thus the functions $S_n(\theta)$ are formed by applying a particular method of summation to the Fourier sums for $f(\theta)$. Rewritten in the new notation we have

$$(3.1.8) \quad |1-d_{n,\nu}| \leq g\, \nu^{k+1}/n^{k+1} .$$

For the sake of uniformity we shall introduce a coefficient $d_{n,o} = 1$ and make use of this inequality in subsequent evaluations.

The results of this section, including the precise form of $S_n(\theta)$ are due to Jackson [1911, 1912, 1914]. Theorem 3.1.1 is established by Jackson [1930] by a different method.

§3.2. DIRECT THEOREMS. The results of §3.1 on degree of approximation in the real domain lead directly to corresponding results in the domain of the complex variable. A fundamental theorem in the study of Problem α is

THEOREM 3.2.1. Let E, with boundary C, be a closed limited set in the z-plane bounded by the mutually exterior analytic Jordan curves $C_1, C_2, \ldots, C_\lambda$; let the functions $f_1(z), f_2(z), \ldots, f_\lambda(z)$ belong to the class $L(k,\alpha)$ on $C_1, C_2, \ldots, C_\lambda$ respectively, and let $f(z) \equiv f_\nu(z)$ for z on or interior to C_ν, $\nu = 1,2, \ldots, \lambda$. Then there exist polynomials $p_n(z)$ such that

$$(3.2.1) \quad |f(z) - p_n(z)| \leq M/n^{k+\alpha} , \quad z \text{ on } E.$$

We take $\lambda = 1$ and consider first the unit circle. Here $z = e^{i\theta}$ and the function $f(z) = f(e^{i\theta})$ is a periodic function of period 2π of the real variable θ. Furthermore we know from the results of §1.2 that $f(e^{i\theta})$ as a function of θ satisfies the hypothesis of Theorem 3.1.1 and thus there exists a trigonometric sum $S_n(\theta)$ of order n in θ such that

$$|f(e^{i\theta}) - S_n(\theta)| \leq M_1/n^{k+\alpha}$$

for all θ. Also $S_n(\theta)$ is a summation of the Fourier development of $f(e^{i\theta})$.

But the Fourier development of $f(e^{i\theta})$ as a function of θ coincides with the Taylor development of $f(z)$ as a function of z on $|z| = 1$. We have

$$f(e^{i\theta}) = \frac{a_0}{2} + \sum_{\nu=1}^{\infty} (a_\nu \cos \nu\theta + b_\nu \sin \nu\theta),$$

$$a_\nu = \frac{1}{\pi} \int_{-\pi}^{\pi} f(e^{i\theta}) \cos \nu\theta \, d\theta, \quad b = \frac{1}{\pi} \int_{-\pi}^{\pi} f(e^{i\theta}) \sin \nu\theta \, d\theta.$$

Applying Euler's formulae to $\cos \nu\theta$, $\sin \nu\theta$, we may write the n-th partial sum of this series in the form

$$t_n(\theta) = \frac{a_0}{2} + \sum_{\nu=1}^{n} [\tfrac{1}{2}(a_\nu - ib_\nu)e^{i\nu\theta} + \tfrac{1}{2}(a_\nu + ib_\nu)e^{-i\nu\theta}].$$

We have

$$\tfrac{1}{2}(a_\nu - ib_\nu) = \frac{1}{2\pi}\int_{-\pi}^{\pi} f(e^{i\theta})(\cos \nu\theta - i \sin \nu\theta)\, d\theta$$

$$= \frac{1}{2\pi i} \int_{|z|=1} \frac{f(z)\,dz}{z^{\nu+1}},$$

$$\tfrac{1}{2}(a_\nu + ib_\nu) = \frac{1}{2\pi i} \int_{|z|=1} f(z)\, z^{\nu-1}\, dz;$$

in this case $(a_\nu + ib_\nu) = 0$ since $f(z)$ is analytic in $|z| < 1$. Thus if we set $2c_\nu = a_\nu - ib_\nu$ we see that $t_n(\theta)$ is the n-th partial sum of the Taylor development

$$\sum_{\nu=0}^{\infty} c_\nu z^\nu .$$

Hence $S_n(\theta)$ is a polynomial of degree n in z. An application of the principle of the maximum yields (3.2.1) and the theorem is established for E the unit circle.

For future reference we observe here that if

$$(3.2.2) \quad f(z) = \sum_{\nu=0}^{\infty} a_\nu z^\nu, \quad |z| \leq 1,$$

then we have

$$(3.2.3) \quad |f(z) - \sum_{\nu=0}^{n} d_{n,\nu}\, a_\nu z^\nu| \leq M/n^{k+\alpha}, \quad |z| \leq 1,$$

where $d_{n,\nu}$ is the summation coefficient of order k investigated in §3.1.

We proceed now to the general case making use of

the above result for the unit circle by conformal mapping. Let $w = \Omega_\nu(z)$ map the interior of C_ν conformally on the interior of $\gamma : |w| = 1$ and denote by $z = \chi_\nu(w)$ the inverse mapping function. Since the curves C_ν, $\nu = 1, 2, \ldots, \lambda$, are analytic and mutually exterior there exists a number $R > 1$ such that the curves C'_ν : $|\Omega_\nu(z)| = R$ are analytic and mutually exterior and such that the functions $\Omega(z)$ are analytic and schlicht in limited simply connected regions containing the C'_ν respectively. By the analyticity of C_ν it is clear that $f_\nu(\chi_\nu(w))$ belongs to the class $L(k, \alpha)$ on γ; therefore for each ν there exist polynomials $P_{\nu,m}(w) = F_{\nu,m}(z)$ in w of respective degrees m such that

(3.2.4) $|f_\nu(z) - F_{\nu,m}(z)| \leq M_1/m^{k+\alpha}$, z on $\overline{C}_\nu$.

The functions $F_{\nu,m}(z)$ are analytic in the limited simply connected region bounded by C'_ν and are uniformly bounded on C_ν by virtue of (3.2.4). Furthermore $F_{\nu,m}(z)$ is a polynomial of degree m in w and hence by Theorem 2.1.3 we have

(3.2.5) $|F_{\nu,m}(z)| \leq M_2 R^m$ for $|w| = R$ or $|\Omega_\nu(z)| = R$.

Consequently $|F_{\nu,m}(z)| \leq M_2 R^m$ for z interior to C'_ν, which contains C_ν. We define $F_m(z) \equiv F_{\nu,m}(z)$, z on or interior to C'_ν, $\nu = 1, 2, \ldots, \lambda$.

Now let $w = \phi(z)$ have the usual meaning (§1.1). Let $\mu > 1$ be a number such that $\Gamma : |\phi(z)| = \mu$ consists of λ analytic Jordan curves lying respectively interior to the curves C'_ν and containing the curves C_ν in their respective interiors; there exists such a number by virtue of the analyticity of each C_ν (compare the discussion of $d(C, C_\rho)$ in §2.1). We also select numbers μ_1, $1 < \mu_1 < \mu$ and r, $\mu_1/\mu < r < 1$. The λ components of $\Gamma : |\phi(z)| = \mu_1$ lie interior to the respective components of Γ and contain the curves C_ν in their respective interiors. Let $P_n(z)$ be the polynomial of degree n which interpolates to $F_m(z)$ in $n + 1$ equally distributed points on C; then we have in the notation of §2.5

$$F_m(z) - P_n(z) = \frac{1}{2\pi i} \int_\Gamma \frac{\omega_n(z)}{\omega_n(t)} \frac{F_m(t)dt}{(t - z)}, \quad z \text{ on } \Gamma_1 ,$$

where $|F_m(t)| \leq M_2 R^m$ by (3.2.5), and $|\omega_n(z)/\omega_n(t)| \leq M_3(\mu_1/\mu)^n$ by Theorem 2.5.7. Hence

$$|F_m(z) - P_n(z)| \leq \frac{M_4 R^m}{d} \left(\frac{\mu_1}{\mu}\right)^n ,$$

where d is $d(\Gamma_1, \Gamma)$. Now let $n = qm$, where q is a positive integer such that $r_1 = r^q R < 1$, whence we have since $\mu_1/\mu < r$

$$(3.2.6) \quad |F_m(z) - P_{qm}(z)| \leq M_5 R^m r^{qm} = M_5 (r^q R)^m$$

$$= M_5 r_1^m, \quad z \text{ on } \Gamma_1,$$

where M_5 is independent of m and z for m sufficiently large. By the principle of the maximum inequality (3.2.6) is valid for z interior to Γ_1 and hence on E; thus a combination of inequalities (3.2.4) and (3.2.6) yields

$$|f(z) - P_{qm}(z)| \leq \frac{M_1}{m^{k+\alpha}} + M_5 r_1^m, \quad z \text{ on E},$$

$$\leq \frac{M_6}{m^{k+\alpha}}, \quad z \text{ on E},$$

since $r_1 < 1$, for m sufficiently large. We now define the polynomials $p_n(z)$ in the statement of the theorem as follows:

$$p_n(z) \equiv 0, \quad (n = 1,2, \ldots, q-1);$$

$$p_{qm'+h}(z) \equiv P_{qm'}(z), \quad (h=0,1,2, \ldots, q-1; \; m' = 1,2, \ldots).$$

Since there exist M_1 and M_2 such that

$$\frac{M_1}{(m')^{k+\alpha}} \leq \frac{M_2}{(qm'+h)^{k+\alpha}}$$

for m' sufficiently large, inequality (3.2.1) follows. The proof of Theorem 3.2.1 is complete.

Although the line segment is not an analytic Jordan curve we have an entirely analogous result.

THEOREM 3.2.2. Let $f(z)$ belong to the class $L(k, \alpha)$ on the segment $-1 \leqq z \leqq 1$. Then there exist polynomials $p_n(z)$, $n = 1, 2, \ldots$, such that

(3.2.7) $|f(z) - p_n(z)| \leqq \dfrac{M}{n^{k+\alpha}}$, $-1 \leqq z \leqq 1$.

Since any line segment in the z-plane is transformed into the segment $-1 \leqq w \leqq 1$ in the w-plane by a translation, stretching, and rotation, and under these transformations a polynomial of degree n is invariant, that is, remains a polynomial of degree n, the above theorem is valid for any line segment.

We refer the proof back to Theorem 3.1.1 by setting $z = \cos \theta$; thus $f(z)$ becomes a periodic function of θ of period 2π

$$f(z) = f(\cos \theta) = g(\theta),$$

where $g(\theta)$ is an even function of θ. A deeper interpretation of this transformation is obtained through consideration of the conformal mapping of the exterior of the segment w $|w| > 1$, the details of which are explained in § 8.2. Let $k = 0$, then we have by virtue of the well known properties of $\cos \theta$

$$|g(\theta_1) - g(\theta_2)| \leqq L|\theta_1 - \theta_2|^{\alpha}.$$

Hence by Theorem 3.1.1 a trigonometric sum $S_n(\theta)$ of order n exists such that

$$|g(\theta) - S_n(\theta)| \leqq ML/n^{\alpha}$$

for all θ. Now we use the fact that $g(\theta)$ is an even function of θ; we set

$$C_n(\theta) = \frac{1}{2} [S_n(\theta) + S_n(-\theta)],$$

which is a cosine sum of order n and hence a polynomial $p_n(z)$ of degree n in z; then by virtue of the fact that

$$g(\theta) = \frac{1}{2}\,[g(\theta) + g(-\theta)],$$

we have

$$\begin{aligned}\left|f(z) - P_n(z)\right| &= |g(\theta) - C_n(\theta)| \\ &= \frac{1}{2}\,\left|[g(\theta)+g(-\theta)]-[S_n(\theta)+S_n(-\theta)]\right| \\ &\leq \frac{ML}{n^\alpha}, \quad -1 \leq z \leq 1,\end{aligned}$$

which is (3.2.7) for k = 0.

Now suppose k = 1; by the above discussion a polynomial $P_n'(z)$ exists such that

$$\left|f'(z) - P_n'(z)\right| \leq \frac{ML}{n^\alpha}, \quad -1 \leq z \leq 1.$$

Let

$$\int_1^z P_n'(z)\,dz = P_n(z), \quad f(z) - P_n(z) = r_n(z);$$

since $|r_n'(z)| \leq ML/n^\alpha$, we have

$$\left|r_n(z_1) - r_n(z_2)\right| \leq \frac{ML}{n^\alpha}\,|z_1 - z_2|,$$

thus there exists a polynomial $Q_n(z)$ of degree n in z such that

$$\left|r_n(z) - Q_n(z)\right| \leq \left(\frac{ML}{n^\alpha}\right)\frac{M_1}{n};$$

by putting $p_n(z) = P_n(z) + Q_n(z)$ we have

$$\left|f(z) - p_n(z)\right| \leq \frac{M_2}{n^{1+\alpha}} \quad -1 \leq z \leq 1.$$

The inequality for arbitrary k follows by applying the above method a suitable number of times. This completes the proof of Theorem 3.2.2 and illustrates an important method in the theory of approximation; compare Chapter VIII.

Theorem 3.2.1 is due to Curtiss [1936]; the statement here is slightly more general since Curtiss assumes that $f^{(k)}(z)$ as well as $f(z)$ is continuous on E [compare Walsh and Sewell, 1937a]. From Theorems 1.2.12 and 1.2.14 it is clear that the continuity of $f(z)$ on E and the Lipschitz condition of $f^{(k)}(z)$ on C imply the continuity of $f^{(k)}(z)$ on E, and in fact the same Lipschitz condition of $f^{(k)}(z)$ on E. The curves $C_1, C_2, \ldots, C_\lambda$ are analytic Jordan curves and hence inequality (1.2.22) is obviously satisfied. Theorem 3.2.2 is due to Jackson [1930].

§3.3. THE FABER POLYNOMIALS. In Theorem 3.2.1 we proved an existence theorem, that is, we showed that polynomials exist which converge with a certain degree; in the present section we actually exhibit a set of polynomials with that degree of convergence provided the set E is bounded by a single analytic Jordan curve. An important property of the polynomials which we study here is their close relation to the Taylor development of which they are a generalization; this relation enables us to use the classical results on the Fourier development and in particular Theorem 3.1.1. Faber [1903, 1920] first studied these polynomials and the development of a function in a sequence of these polynomials; he obtained interesting results on geometric degree of convergence. We study here a summation of the Faber development, in fact the summation employed by Jackson in the proof of Theorem 3.1.1. This section might well be titled: "The Jackson summation of the Faber development".

The Faber polynomials are defined through the mapping function. Let C be an analytic Jordan curve, and let

$$(3.3.1)\quad z = \psi(w) = cw + \mathfrak{P}\left(\frac{1}{w}\right), \quad c > 0,$$

map the exterior of C in the z-plane onto the exterior of $\gamma: |w| = 1$ so that the points at infinity in the two planes correspond to each other, and so that the constant c is real and positive. In (3.3.1) the function $\mathfrak{P}(1/w)$ is a power series in $1/w$ which converges for $|w| \geq r$, where r is some number less than unity, by virtue of the analyticity of C; the constant c is the capacity of C or of $\overline{C}$. Let f(z) be analytic in C and continuous on $\overline{C}$; then by the Cauchy integral formula we have

$$f(z) = \frac{1}{2\pi i} \int_C \frac{f(t)}{t-z}\, dt,$$

which in the w-plane becomes

$$(3.3.2) \quad f(z) = \frac{1}{2\pi i} \int_\gamma \frac{f(\psi(w))}{\psi(w) - z} \frac{d\psi(w)}{dw}\, dw.$$

Thus Faber is led to study the function under the integral sign in (3.3.2); in this connection he shows that

$$(3.3.3) \quad \frac{w\psi'(w)}{\psi(w)-z} = -1 + \sum_{\nu=0}^{\infty} F_{\nu+1}(z) w^{-\nu+1},$$

where $F_\nu(z)$ is a polynomial of degree ν in z. The Cauchy coefficient theorem together with the analyticity of $\psi(w)$ in $|w| \geq r$ yields the inequality

$$(3.3.4) \quad |F_{\nu+1}(z)| \leq G_1 r^{\nu+1}, \; r < 1, \; z \text{ on } C_r$$

where G_1 is a constant depending on z; the curve C_r is the map of $|w| = r < 1$ under the transformation $z = \psi(w)$. The series $\sum_{\nu=0}^{\infty} F_{\nu+1}(z) w^{-\nu+1}$ represents a function which is uniformly bounded for all $|w| \geq r$ and for all z on or exterior to C_r, hence the constant G_1 is uniformly bounded for all z on or exterior to C_r. Furthermore

$$\frac{w\psi'(w)}{\psi(w) - \psi(w')} = -1 + \sum_{\nu=0}^{\infty} (-w'^{\nu+1} + \mathfrak{P}_{\nu+1}(\tfrac{1}{w'}))w^{-\nu+1}, \; |w'| > |w|,$$

where $\mathfrak{P}_{\nu+1}(1/w')$ is a power series in $1/w'$; hence

$$F_\nu(z) = -w^\nu + \mathfrak{P}_\nu(\tfrac{1}{w}), \; |w| > r < 1.$$

Thus applying the Cauchy coefficient theorem again we see that

$$(3.3.5) \quad |\mathfrak{P}_\nu(\tfrac{1}{w})| \leq G_2 r^\nu, \; |w| > r,$$

where G_2 is a constant; hence

$$(3.3.6)\quad F_\nu(z) = w^\nu\,(-1 + \theta_\nu(z)\,G),$$

where G is a constant, and $|\theta_\nu(z)| < (r/\rho)^\nu$ for all z on C_ρ, $\rho > r$. It follows by the Cauchy integral formula that f(z) can be developed in a series

$$(3.3.7)\quad f(z) = \sum_{\nu=0}^{\infty} a_\nu F_\nu(z),$$
$$a_\nu = -\frac{1}{2\pi i}\int_\gamma f(\psi(w))\frac{dw}{w^{\nu+1}},$$

which converges to f(z) interior to C. In fact if we assume that f(z) belongs to the class $L(0,\alpha)$ on C then the series converges to f(z) on $\overline{C}$ by the well known properties of the Fourier and Taylor developments. This means that we have in the w-plane

$$(3.3.8)\quad \begin{aligned} f(z) &= \sum_{\nu=0}^{\infty} a_\nu\left(-w^\nu + \mathfrak{P}_\nu\left(\tfrac{1}{w}\right)\right) \\ &= -\sum_{\nu=0}^{\infty} a_\nu w^\nu + \sum_{\nu=0}^{\infty} a_\nu\,\mathfrak{P}_\nu\left(\tfrac{1}{w}\right), \\ & r \leq |w| \leq 1, \end{aligned}$$

or a Taylor series and a power series.

We have shown by (3.2.3) that the Jackson summation of the Taylor development gives the desired degree of convergence for the unit circle; thus it is natural to consider the same summation of the Faber development for an analytic Jordan curve; in fact we prove

THEOREM 3.2.1. Let C be an analytic Jordan curve and let f(z) belong to the class $L(k,\alpha)$ on C. Then we have

$$(3.3.9)\quad \left|f(z) - \sum_{\nu=0}^{n} d_{n,\nu}\,a_\nu F_\nu(z)\right| \leq M/n^{k+\alpha},\quad z \text{ on } \overline{C},$$

where $\sum_0^n a_\nu F_\nu(z)$ is the sum of the first n+1 terms of the development (3.3.7) of f(z) in the Faber polynomials belonging to C, and $d_{n.\nu}$ is the

Jackson summation coefficient of order k.

From (3.3.8) we have

$$(3.3.10)\quad f(z) = -\sum_{\nu=0}^{\infty} a_\nu w^\nu + \sum_{\nu=1}^{\infty} c_\nu / w^\nu, \quad r \leq |w| \leq 1.$$

Now we consider

$$f(z) - \sum_{\nu=0}^{n} d_{n,\nu}\, a_\nu F_\nu(z) = f(\psi(w))$$

$$- \sum_{\nu=0}^{n} d_{n,\nu}\, a_\nu \left(-w^\nu + \mathfrak{P}_\nu\left(\tfrac{1}{w}\right)\right)$$

$$= f(\psi(w)) + \sum_{\nu=0}^{n} d_{n,\nu}\, a_\nu w^\nu - \sum_{\nu=1}^{\infty} c_\nu / w^\nu$$

$$- \sum_{\nu=1}^{n} d_{n,\nu}\, a_\nu \mathfrak{P}_\nu\left(\tfrac{1}{w}\right) + \sum_{\nu=1}^{\infty} c_\nu / w^\nu$$

$$(3.3.11)\quad = f(\psi(w)) - \sum_{\nu=1}^{\infty} c_\nu / w^\nu + \sum_{\nu=0}^{n} d_{n,\nu}\, a_\nu w^\nu$$

$$+ \sum_{\nu=1}^{n} (1 - d_{n,\nu}) a_\nu \mathfrak{P}_\nu\left(\tfrac{1}{w}\right) + \sum_{\nu=n+1}^{\infty} a_\nu \mathfrak{P}_\nu\left(\tfrac{1}{w}\right),$$

$$r \leq |w| \leq 1;$$

since we have uniform convergence these transformations are permissible. It is clear from (3.3.10) that $-\sum_{0}^{n} d_{n,\nu}\, a_\nu w^\nu$ is the sum of the first n + 1 terms of the Jackson summation of the Fourier development of the function $f(\psi(w)) - \sum_{1}^{\infty} c_\nu / w^\nu$. The function $f^{(k)}(\psi(w))$ satisfies a Lipschitz condition of order α on $|w| = 1$ by virtue of the analyticity of C and the fact that f(z) belongs to the class $L(k, \alpha)$ on C; also $\sum_{1}^{\infty} c_\nu / w^\nu$ is analytic on $|w| = 1$. Hence we know from (3.2.3) that

$$(3.3.12)\quad \left| f(\psi(w)) - \sum_{1}^{\infty} \frac{c_\nu}{w^\nu} + \sum_{0}^{n} d_{n,\nu} a_\nu w^\nu \right|$$

$$\leq M_1 / n^{k+\alpha}, \quad |w| = 1,$$

which is valid in fact for $r \leq |w| \leq 1$ by virtue of analyticity and continuity. Also by (3.3.7) we see that

$$|a_\nu| \leq M_2 / 2\pi;$$

thus by virtue of (3.3.5)

$$(3.3.13) \quad \left| \sum_{n+1}^{\infty} a_\nu \mathfrak{p}_\nu\left(\frac{1}{w}\right) \right| \leqq \sum_{n+1}^{\infty} |a_\nu| \left| \mathfrak{p}_\nu\left(\frac{1}{w}\right) \right|$$
$$\leqq M_3 \sum_{n+1}^{\infty} r^\nu, \quad r < 1,$$
$$\leqq M_4/n^{k+\alpha},$$

for suitably chosen M_4. Furthermore by the inequality (3.1.8):

$$|1 - d_{n,\nu}| \leqq M_5 \nu^{k+1}/n^{k+1},$$

we have

$$\left| \sum_{\nu=1}^{n} (1-d_{n,\nu}) a_\nu \mathfrak{p}_\nu\left(\frac{1}{w}\right) \right| \leqq M_6 \sum_{1}^{n} \nu^{k+1} r^\nu / n^{k+1}, \quad r < 1,$$
$$\leqq \frac{M_7}{n^{k+1}} \sum_{1}^{n} \nu^{k+1} r^\nu;$$

but the series on the right converges and we obtain

$$(3.3.14) \quad \left| \sum_{\nu=1}^{n} (1 - d_{n,\nu}) a_\nu \mathfrak{p}_\nu\left(\frac{1}{w}\right) \right| \leqq M_8/n^{k+\alpha}.$$

From equation (3.3.11) and inequalities (3.3.12), (3.3.13), and (3.3.14) we see that inequality (3.3.9) is valid for z on C, hence for z on $\overline{C}$ by the principle of the maximum. The proof of the theorem is complete.

The Faber development might well be considered an extension of the Taylor development, and results on degree of approximation of the development itself are included in the exercises at the end of this chapter. We also consider this development further in connection with Problem β. Theorem 3.3.1 is due to the author [1939].

§3.4. INDIRECT THEOREMS. In the preceding sections of this chapter we assumed conditions on the function and studied the degree of convergence of various sequences of approximating polynomials; theorems in the converse direction are considered in the present section. The results are stated for a single closed Jordan region but they obviously hold for any finite number of such regions.

Our main result is

THEOREM 3.4.1. Let C be a Jordan curve or arc of Type t and let $f(z)$ be defined in $\overline{C}$. For each n, n = 1,2,..., let a polynomial $P_n(z)$ exist such that

$$(3.4.1) \quad |f(z) - P_n(z)| \leq M/n^{(k+\alpha)t}, \quad z \text{ in } \overline{C}.$$

Then $f^{(k)}(z)$ exists on $\overline{C}$ and satisfies the condition

$$(3.4.2) \quad |f^{(k)}(z_1) - f^{(k)}(z_2)| \leq$$

$$\leq L|z_1-z_2|^{\alpha} \left|\log |z_1-z_2|\right|^{\beta}, \quad z_1, z_2 \text{ in } \overline{C}$$

where $\beta = 1$ if $\alpha = 1$ and $\beta = 0$ if $\alpha < 1$, L is a constant independent of z_1 and z_2, and $|z_1-z_2|$ is sufficiently small.

This theorem states that for C an analytic Jordan curve (t = 1) and $0 < \alpha < 1$ the function $f(z)$ belongs to the class $L(k, \alpha)$ and hence is an exact converse of Theorem 3.2.1; however, for $\alpha = 1$ the function belongs to the class Log (k,1) rather than to the class L(k,1). The following example shows that for $\alpha = 1$ an exact converse of Theorem 3.2.1 is impossible. Let C be the unit circle and let

$$f(z) = \sum_{\nu=2}^{\infty} \frac{z^{\nu}}{\nu(\nu-1)}, \quad |z| \leq 1;$$

then

$$\left|f(z) - \sum_{\nu=2}^{n} \frac{z}{\nu(\nu-1)}\right| \leq \sum_{n+1}^{\infty} \frac{1}{\nu(\nu-1)}$$

$$= \sum_{n+1}^{\infty} \left(\frac{1}{\nu-1} - \frac{1}{\nu}\right) = \frac{1}{n}, \quad n = 2, 3, \ldots, |z| \leq 1.$$

But $f(z)$ does not satisfy a Lipschitz condition of order 1 on $|z| = 1$ since the derivative $f'(z) = -\log(1-z)$ becomes infinite as z approaches 1. We have taken k = 0 but the extension to arbitrary positive integral k follows immediately by integrating f(z) a suitable number of times.

In order to be able to give a complete proof of Theorem 3.4.1 and also to clarify the method of proof we establish first the following theorem:

THEOREM 3.4.2. Let C be a Jordan arc or curve of Type t and let f(z) be defined on $\overline{C} \equiv E$. For each n, n = 1,2,..., let a polynomial $P_n(z)$ exist such that

(3.4.3) $|f(z) - P_n(z)| \leq M/n^{\beta t}$, z on E, $\beta > 1$.

Then f(z) is analytic interior to C, continuous in $E = \overline{C}$, and f'(z) exists in E and we have

(3.4.4) $|f'(z) - P_n'(z)| \leq M_1/n^{(\beta-1)t}$, z on E.

The analyticity and continuity follow directly from inequality (3.4.3). Also from (3.4.3) we have

(3.4.5) $|f(z) - P_{n+1}(z)| \leq M/(n+1)^{\beta t}$, z on E,

and from (3.4.3) and (3.4.5) it is clear that

$$|P_{n+1}(z) - P_n(z)| \leq 2M/n^{\beta t}, \quad z \text{ on } E.$$

Thus by Theorem 2.1.4 we have

(3.4.6) $|P_{n+1}'(z) - P_n'(z)| \leq 2M(n+1)^t/n^{\beta t} \leq M_1/n^{(\beta-1)t}$,

for z on E. If we write

(3.4.7) $f(z) = P_1(z) + [P_2(z) - P_1(z)] + \dots$

$$+[P_{n+1}(z) - P_n(z)] + \dots,$$

by (3.4.6) the series of derivatives converges uniformly on E for $\beta > 2$. Thus we have

$$g(z) = P_1'(z) + [P_2'(z) - P_1'(z)] + \dots;$$

since the series of derivatives converges uniformly on E we can integrate the series g(z) term by term between the limits z_0 and z

$$\int_{z_0}^{z} g(z)dz = [P_1(z)-P_1(z_0)]+[P_2(z)-P_1(z)-(P_2(z_0)-P_1(z_0))]+\ldots$$
$$= f(z) - f(z_0).$$

Since the derivative of the left hand member obviously exists the derivative $f'(z)$ of the right hand member exists and we have

$$f'(z) = g(z) = P_1'(z) + [P_2'(z) - P_1'(z)] + \ldots$$

For $\beta > 1$ we consider a subsequence of (3.4.7). Let n be arbitrary but fixed and choose an integer m such that

$$2^{m-1} \leq n < 2^m;$$

this is possible no matter what the integer n may be. Now from (3.4.7) we may write

$$(3.4.8)\quad f(z)-P_n(z) = [P_{2^m}(z)-P_n(z)]$$
$$+ [P_{2^{m+1}}(z) - P_{2^m}(z)] + \ldots ;$$

from inequality (3.4.3) we have

$$(3.4.9)\quad \left|f(z) - P_{2^m}(z)\right| \leq M/2^{m\beta t}, \quad z \text{ on } E,$$

$$(3.4.10)\quad \left|f(z) - P_{2^{m+1}}(z)\right| \leq M/2^{(m+1)\beta t}, \quad z \text{ on } E.$$

A combination of (3.4.3) and (3.4.9) yields

$$\left|P_{2^m}(z) - P_n(z)\right| \leq 2M/n^{\beta t}, \quad z \text{ on } E;$$

a combination of (3.4.9) and (3.4.10) yields

$$(3.4.11)\quad \left|P_{2^{m+1}}(z) - P_{2^m}(z)\right| \leq 2M/2^{m\beta t}, \quad z \text{ on } E.$$

Now applying Theorem 2.1.4 we have

$$\left|P'_{2^m}(z) - P'_n(z)\right| \leq M_1 2^{mt}/n^{\beta t}, \quad z \text{ on } E,$$

$$(3.4.12)\quad \left|P'_{2^{m+1}}(z) - P'_{2^m}(z)\right| \leqq M_1 2^{(m+1)t}/2^{m\beta t}, \quad z \text{ on } E,$$

where M_1 is a constant independent of n and of m. Using these evaluations we see that in the series

$$(3.4.13)\quad [P'_{2^m}(z) - P'_n(z)] + [P'_{2^{m+1}}(z) - P'_{2^m}(z)] + \cdots$$

the first term is in modulus not greater than

$$\frac{M_1 2^{mt}}{2^{(m-1)\beta t}} = \frac{2^{\beta t}M_1}{2^{(\beta-1)tm}};$$

the second term is in modulus not greater than

$$\frac{2^{\beta t}M_1}{2^{(\beta-1)tm}} = \frac{2^{\beta t}M_1}{2^{(\beta-1)t(m+1)}}, \text{ etc.}$$

Hence (3.4.13) is in modulus not greater than

$$\frac{2^{\beta t}M_1}{2^{(\beta-1)tm}} + \frac{2^{\beta t}M_1}{2^{(\beta-1)t(m+1)}} + \frac{2^{\beta t}M_1}{2^{(\beta-1)t(m+2)}} + \cdots$$

which equals

$$\frac{2^{\beta t}M_1}{2^{(\beta-1)tm}} \left\{ 1 + \frac{1}{2^{(\beta-1)t}} + \frac{1}{[2^{(\beta-1)t}]^2} + \cdots \right.$$

$$\left. + \frac{1}{[2^{(\beta-1)t}]} + \cdots \right\}.$$

Since we assume $\beta > 1$ the series in braces converges and in fact to a constant depending on β and t but independent of n and m. Thus (3.4.13) is in modulus less than

$$\frac{M_2}{(2^m)^{(\beta-1)t}} < \frac{M_2}{n^{(\beta-1)t}},$$

where M_2 is a constant independent of n. Thus by (3.4.8) we see that (3.4.4) is satisfied and the proof of the theorem is complete.

We are now ready to prove Theorem 3.4.1. We consider

first the case $k = 0$, $\alpha < 1$. As above we write

$$f(z) = P_2(z) + \sum_{n=1}^{\infty} [P_{2^{m+1}}(z) - P_{2^m}(z)],$$

hence

$$\begin{aligned} f(z_1)-f(z_2) = P_2(z_1) &+ \sum_{m=1}^{\mu} [P_{2^{m+1}}(z_1) - P_{2^m}(z_1)] \\ &+ \sum_{m=\mu+1}^{\infty} [P_{2^{m+1}}(z_1) - P_{2^m}(z_1)] \\ &- P_2(z_2) - \sum_{m=1}^{\mu} [P_{2^{m+1}}(z_2)-P_{2^m}(z_2)] \\ &- \sum_{m=\mu+1}^{\infty} [P_{2^{m+1}}(z_2) - P_{2^m}(z_2)], \end{aligned}$$

where μ is a positive integer to be determined later. It is clear that $|P_2(z_1)-P_2(z_2)| \leqq L_1|z_1-z_2|$, where L_1 is a constant independent of n, μ, z_1, and z_2. By virtue of (3.4.12) and the fact that the arc and chord of C are infinitesimals of the same order we have (See(2.5.34),ff.)

$$\begin{aligned} &\left| \sum_{m=1}^{\mu} [P_{2^{m+1}}(z_1)-P_{2^m}(z_1)] - \sum_{m=1}^{\mu} [P_{2^{m+1}}(z_2) - P_{2^m}(z_2)]\right| \\ &\leqq \sum_{m=1}^{\mu} \frac{2M_3 2^{(m+1)t}}{2^{m\alpha t}} \cdot |z_1 - z_2| \\ &\leqq |z_1-z_2| \left[2M_3 2^t \sum_{m=1}^{\mu} 2^{mt(1-\alpha)} \right] \\ &= |z_1-z_2| \; 2M_3 2^t \left[2^{\mu t(1-\alpha)} + \frac{2^{\mu t(1-\alpha)}}{2^{t(1-\alpha)}} + \cdots + \frac{2^{\mu t(1-\alpha)}}{(2^{t(1-\alpha)})^{\mu-1}} \right] \\ &= |z_1-z_2| 2M_3 2^t 2^{\mu t(1-\alpha)} \\ &\quad \left[1 + \frac{1}{2^{t(1-\alpha)}} + \cdots + \frac{1}{(2^{t(1-\alpha)})^{\mu-1}} \right] \\ &< M_4 2^{\mu t(1-\alpha)} |z_1 - z_2|; \end{aligned}$$

since we have $\alpha < 1$ the expression in brackets is a partial sum of an absolutely convergent series and M_4 is is independent of μ.

By (3.4.11) it is clear that

$$\left| \sum_{m=\mu+1}^{\infty} [P_{2^{m+1}}(z_1)-P_{2^m}(z_1)] - \sum_{m=\mu+1}^{\infty} [P_{2^{m+1}}(z_2)-P_{2^m}(z_2)] \right|$$

$$\leq 2 \sum_{m=\mu+1}^{\infty} 2M/2^{m\alpha t}$$

$$< M_5/2^{\mu\alpha t},$$

where M_5 is a constant independent of μ.

Thus

$$|f(z_1)-f(z_2)| \leq L_1|z_1-z_2|+M_4 2^{\mu t(1-\alpha)}|z_1-z_2|+ M_5 \frac{1}{2^{\mu\alpha t}}.$$

It is sufficient to establish (3.4.2) for $|z_1-z_2|$ small. Given $|z_1-z_2|$ arbitrary but small we choose μ so that

$$|z_1-z_2| < \frac{1}{2^{\mu t}} \leq 2|z_1-z_2|;$$

such an integer μ exists because these inequalities simply assert that given any number $1/|z_1-z_2|$ sufficiently large, then there exists an integer μ such that

$$2^{\mu} < \frac{1}{|z_1-z_2|^{1/t}} \leq 2^{\mu+1}.$$

From these inequalities we have

$$2^{\mu t(1-\alpha)} < \frac{1}{|z_1-z_2|^{t(1-\alpha)/t}} = |z_1-z_2|^{\alpha-1},$$

$$\frac{1}{(2^{\mu+1})^{\alpha t}} \leq (|z_1-z_2|^{1/t})^{\alpha t} = |z_1 - z_2|^{\alpha},$$

or

$$\frac{1}{2^{\mu\alpha t}} \leq M_7 |z_1 - z_2|^{\alpha}.$$

Combining these inequalities we have (3.4.2) for $k = 0$, $0 < \alpha < 1$.

If $\alpha = 1$, with k still zero, the above evaluation

does not go through; however we have

$$|f(z_1)-f(z_2)| \leq L_1|z_1-z_2| + M_3\,\mu\,|z_1-z_2| + M_4\,\frac{1}{2^{\mu\alpha t}},$$

and taking the same value for μ we obtain

$$2^{\mu} < \frac{1}{|z_1 - z_2|^{1/t}}$$

$$\mu \log 2 < -\frac{1}{t}\log|z_1-z_2|$$

$$\mu < -\frac{1}{t\log 2}\log|z_1-z_2|.$$

Thus

$$|f(z_1)-f(z_2)| \leq L|z_1-z_2|\,\bigl|\log|z_1-z_2|\bigr|,$$

where L is a constant independent of z_1 and z_2.

Now let k be an arbitrary positive integer. By applying Theorem 3.4.2 to the inequality

$$\left|f(z) - P_n(z)\right| \leq M/n^{(k+\alpha)t}, \quad z \text{ on } E,$$

k times we obtain

$$\left|f^{(k)}(z) - P_n^{(k)}(z)\right| \leq M_1/n^{\alpha t}, \quad z \text{ on } E.$$

Thus we have (3.4.2) for $f^{(k)}(z)$ and the proof of Theorem 3.4.1 is complete

Of course it is possible to carry this investigation much further by using the same methods. Various types of convergence yield various types of continuity; de la Vallée Poussin [1919, Chap. IV] treats similar questions for the real domain in some detail.

It should be observed that (3.4.2) is a property in the small and if $t = 1$ for an arc of C then on any proper sub-arc we have a Lipschitz condition of order α; this gives a proof of the following theorem

THEOREM 3.4.3. Let f(z) be defined in

$-1 \leq z \leq 1$ and let $P_n(z)$ exist such that

$$|f(z)-P_n(z)| \leq M/n^{k+\alpha}, \quad -1 \leq z \leq +1.$$

Then in any interior interval $-1 < a \leq z \leq b < 1$ the derivative $f^{(k)}(z)$ exists and satisfies condition (3.4.2).

We have merely to recall that for the line segment $(-1, +1)$ the value of t is 2 but we have $t = 1$ for any proper interior interval.

Theorem 3.4.1 for C an analytic Jordan curve is due to Walsh and Sewell [1937a] along with the example; for non-analytic curves the author [1938a] established a weaker form; the method of splitting the series at μ is employed by de la Vallée Poussin [1919] in studying functions of a real variable [see also Sewell, 1939a]. Special cases of Theorem 3.4.2 are due to the author [1936,1937]; the method is used by Montel [1919] in the study of generalized derivatives and approximation in the real domain. Theorem 3.4.3 is due to de la Vallée Poussin [1919].

§3.5. THE DERIVATIVE AND INTEGRAL OF A FUNCTION. In Theorem 3.4.2 we established in a particular case a relation between the degree of approximation to a function by a sequence of polynomials and the degree of approximation to the derivative of the function by the derivative of this sequence of polynomials; the present section is devoted to this problem in more generality and to the corresponding problem for integration.

Let C be an analytic Jordan curve. If $f(z)$ belongs to the class $L(k, \alpha)$ $k \geq 1$, on C then by Theorem 3.2.1 there exist polynomials $P_n(z)$ such that

$$(3.5.1) \qquad |f(z) - P_n(z)| \leq M/n^{k+\alpha}, \quad z \text{ on } \overline{C};$$

furthermore $f'(z)$ belongs to the class $L(k-1, \alpha)$ on C and hence by the same theorem polynomials $p_n(z)$ exist such that

$$|f'(z) - p_n(z)| \leq M/n^{k-1+\alpha}, \quad z \text{ on } \overline{C}.$$

If we set

$$F(z) = \int_{z_0}^{z} f(z)\, dz,$$

where z_0 is fixed in $\overline{C}$ and the path of integration is along permissible arcs in $\overline{C}$, then $F(z)$ belongs to the class $L(k+1, \alpha)$ on C and polynomials $q_n(z)$ exist such that

$$|F(z) - q_n(z)| \leq M/n^{k+1+\alpha}, \quad z \text{ on } \overline{C}.$$

On the other hand if $f(z)$ is defined on $\overline{C}$ and polynomials $P_n(z)$ exist such that (3.5.1) is valid we know by Theorem 3.4.1 that $f(z)$ belongs to the class $L(k, \alpha)$, $0 < \alpha < 1$, on C.

If the function $f(z)$ belongs to the class $L(k, \alpha)$, $k \geq 1$, on C then all we can say about $f'(z)$ is that it belongs to the class $L(k-1, \alpha)$; thus it follows from the above discussion that Theorem 3.4.2 is the best possible as far as the exponent of n is concerned. In this connection the following example [Walsh and Sewell, 1940] is of interest. Let

$$(3.5.2) \quad f(z) = \sum_{\nu=0}^{\infty} a_\nu z^\nu$$

belong to the class $L(1, \alpha)$ on γ: $|z| = 1$; also suppose that the coefficients a_ν, $\nu = 0, 1, 2, \ldots$, are all real and non-negative numbers. Then we have

$$\max\left[\left|f'(z) - \sum_{\nu=1}^{n} \nu a_\nu z^{\nu-1}\right|, |z| = 1\right] = \sum_{\nu=n+1}^{\infty} \nu a_\nu$$

$$\geq n \sum_{\nu=n+1}^{\infty} a_\nu$$

$$= n \max\left[\left|f(z) - \sum_{\nu=0}^{n} a_\nu z^\nu\right|, |z| = 1\right].$$

Now let us consider the integral $F(z)$. If we denote by $P_{n+1}(z)$ the corresponding indefinite integral of

$P_n(z)$ then we have

$$\left|F(z) - P_{n+1}(z)\right| = \left| \int_{z_0}^{z} [f(z) - P_n(z)]\, dz \right|$$
$$\leq \{ \max_{z \text{ on } C} |f(z) - P_n(z)| \} M,$$

where M is a constant depending on C and z_0. Consequently if $|f(z)-P_n(z)| \leq \epsilon_n$, z on $\overline{C}$, then $|F(z)-P_{n+1}(z)| \leq M\epsilon_n$, z on $\overline{C}$; this is in general all that can be said as the following example shows. Let

$$f(z) = 1 + \frac{1}{2^\beta} + \frac{1}{3^\beta} + \ldots + \frac{1}{n^\beta} + \ldots, \quad \beta > 2.$$

$$P_n(z) = 1 + \frac{1}{2^\beta} + \ldots + \frac{1}{n^\beta},$$

then

$$\left|f(z) - P_n(z)\right| \leq M/n^{\beta-1};$$

by integration we have

$$\left| \int_{z_0}^{z} [f(z) - P_n(z)]\, dz \right| \geq \frac{M}{n^{\beta-1}} \left| \int_{z_0}^{z} dz \right|$$
$$= \frac{M}{n^{\beta-1}} | z - z_0 |,$$

where we can make $|z-z_0|$ arbitrarily large by suitable choice of C.

Since the integral of a function in general possesses higher continuity properties than the function itself the fact that the degree of approximation need not be increased by integration is rather disappointing. On the other hand there are circumstances where a positive conclusion can be drawn; we consider the example (3.5.2) again:

$$\left| \int_0^z [f(z) - \sum_{\nu=0}^{n} a_\nu z^\nu]dz \right| = \left| \int_0^z \sum_{n+1}^{\infty} a_\nu z^\nu dz \right|$$
$$= \left| \sum_{n+1}^{\infty} \frac{a_\nu z^{\nu+1}}{\nu+1} \right| \leq \frac{M}{n} \left| f(z) - \sum_{\nu=0}^{n} a_\nu z^\nu \right|, \quad |z| \leq 1.$$

Here we have increased the exponent of n by unity; this is in general all that can be expected. For suppose f(z) is defined on $\overline{C}$, with t = 1 for simplicity, and suppose $P_n(z)$ exists such that

$$|f(z) - P_n(z)| \leq M_1/n^{\beta}, \quad z \text{ on } \overline{C};$$

furthermore suppose that the value of β is such that there exists at least one point z_1 of $\overline{C}$ for which we have

$$|f(z_1) - P_n(z)_1)| \geq M_2/n^{\beta}.$$

Now let us assume that

$$|F(z) - P_{n+1}(z)| \leq M_3/n^{\delta}, \quad z \text{ on } \overline{C},$$

then by Theorem 3.4.3 we have

$$|F'(z) - P'_{n+1}(z)| = |f(z) - P_n(z)| \leq M_4/n^{\delta-1}, \quad z \text{ on } \overline{C}.$$

This leads us to the conclusion that

$$\frac{M_4}{n^{\delta-1}} \geq \frac{M_2}{n^{\beta}},$$

or that $\delta \leq \beta + 1$ in this case.

§3.6. BEST APPROXIMATION. In our study of approximation up to this point we have been content with showing that certain continuity properties imply a sequence of polynomials converging with a predictable degree of approximation, and conversely; in the present section we consider best approximation and thus in a sense make an appraisal of the quality of our results.

Let f(z) be defined on a set E and let $p_n(z)$ be given such that

$$|f(z) - p_n(z)| \leq \epsilon_n, \quad z \text{ on } E;$$

if f(z) is continuous on E and if

$$\max_{z \text{ on } E} |f(z) - p_n(z)| \leq \max_{z \text{ on } E} |f(z) - q_n(z)|,$$

where $q_n(z)$ is any polynomial of degree n other than

$p_n(z)$, then $p_n(z)$ is a <u>polynomial of best approximation</u> to $f(z)$ on E <u>in the sense of Tchebycheff</u> or a Tchebycheff polynomial of $f(z)$ on E. For the functions and sets which we consider Tonelli [1908] has shown [see, e.g., Walsh, 1935, Chap. XII] that $p_n(z)$ exists and is unique. If

$$\max_{z \text{ on } E} |f(z) - p_n(z)| = \epsilon_n$$

then the infinitesimal ϵ_n is a measure of the <u>best degree of approximation in the sense of Tchebycheff</u> to $f(z)$ on E. The theorems which we have proved in this chapter lead immediately to certain inequalities on ϵ_n.

Let E, with boundary C, be a closed limited set bounded by a finite number of mutually exterior analytic Jordan curves and let $f(z)$ belong to the class $L(k, \alpha)$ on C; then we know from Theorem 3.2.1 that the quantity $\epsilon_n \cdot n^{k+\alpha}$ is bounded. On the other hand if $f(z)$ does not belong to the class $L(k, \alpha')$, $\alpha' > \alpha$ if $0 < \alpha < 1$, on C, or to the class $L(k+1, \alpha')$, $\alpha' > 0$ if $\alpha = 1$, on C, then we know from Theorem 3.4.1 that the quantity $\epsilon_n \cdot n^{k+\alpha'}$ is not bounded. Of course the fact that theorems 3.2.1 and 3.4.1 are exact converses for $\alpha < 1$ indicates that no continuity condition lighter than a Lipschitz condition of order α will lead to an $\epsilon_n < M/n^\alpha$.

We leave to the reader the discussion of the case where C is not an analytic Jordan curve.

§3.7. EXERCISES. 3.7.1. Let C be an analytic Jordan curve and let $f(z)$ belong to the class $L(k, \alpha)$ on C. Then if $\sum_0^\infty a_\nu F_\nu(z)$ is the Faber development of $f(z)$ we have $|a_\nu| \leq M/n^{k+\alpha}$. [Sewell, 1937]

3.7.2. Under the hypothesis of Ex. 3.7.1 we have

$$\left| f(z) - \sum_{\nu=0}^{n} a_\nu F_\nu(z) \right| \leq M \log n/n^{k+\alpha}, \quad z \text{ on } \overline{C}.$$

[Sewell, 1935]

SUGGESTION. In Exs. 3.7.2 and 3.7.3 use the corresponding results on the Taylor (or Fourier) development [compare Jackson, 1930].

3.7.3. Under the hypothesis of Ex. 3.7.1 with k =0, $0 < \alpha < 1$, we have

$$|f(z) - \sigma_n(z)| \leq M/n^{\alpha}, \quad z \text{ on } \overline{C},$$

where $\sigma_n(z)$ is the arithmetic mean of the sum of the first n + 1 terms of the Faber development of f(z). [Walsh and Sewell, 1940]

3.7.4. Let C be a Jordan curve and let f(z) be defined on $\overline{C}$. Let $p_n(z)$ exist such that

$$\left|f(z) - p_n(z)\right| \leq \epsilon_n, \quad z \text{ on } \overline{C},$$

where ϵ_n approaches zero. Then we have

$$\left|f(z) - P_n(z)\right| \leq (n+1)\,\epsilon_n, \quad z \text{ on } \overline{C},$$

where $P_n(z)$ is the polynomial of degree n which interpolates to $f(z) - p_n(z)$ in the n+1 Fekete points of $\overline{C}$.

Note. The Fekete [1926] points of $\overline{C}$ are the points $z_\nu^{(n)}$ of $\overline{C}$ such that the Vandermonde determinant

$$V_n(z_1^{(n)}, z_2^{(n)}, \ldots, z_{n+1}^{(n)}) = \prod_{i<j=1}^{j=n+1} (z_i^{(n)} - z_j^{(n)})$$

is as large in modulus as possible.

SUGGESTION. Use the Lagrange interpolation formula

$$L_n(z) = \sum_{\nu=1}^{n=1} w_\nu \frac{\omega(z)}{(z-z_\nu)\;\omega'(z_\nu)},$$

$$\omega(z) = (z-z_1)(z-z_2)\ldots(z-z_{n+1}),$$

which gives the polynomial $L_n(z)$ of degree n assuming the values w_ν in the points z_ν .

3.7.5. Let C be a Jordan curve such that the function

$$\frac{w\;\psi'(w)}{\psi(w) - \psi(w')} - \frac{w}{w-w'}$$

is bounded on $|w| = 1$ uniformly with respect to w' for

$|w'| = 1$, and is expressible for $|w| > 1$ and $|w'| = 1$ by the Cauchy integral over $|w'| = 1$. Then

$$\left| \frac{F_n(z)}{F_n(t)} \right| \leqq M/\rho^n, \quad z \text{ on } C, \ t \text{ on } C_\rho \ ,$$

where $F_n(z)$ is the Faber polynomial of degree n belonging to C. [Walsh and Sewell, 1940]

3.7.6. What is the relation between the Faber polynomials belonging to C, an analytic Jordan curve, and the Faber polynomials belonging to C_ρ ?

3.7.7. Derive the expression for the Césaro summation (of order unity) of the Taylor development of a function f(z) belonging to the class $L(k, \alpha)$ on $|z| = 1$.

§3.8. OPEN PROBLEMS. There are many interesting problems in connection with the results of this chapter. The proof which we have given of Theorem 3.2.1 is based on corresponding results in the real domain; a purely function-theoretic proof would be an important contribution. Also there is the problem of extending this theorem to more general boundaries; the author [1937b] has obtained some results in this direction but the solution of the problem is still in its infancy; a case of particular interest is that of two or more regions bounded by (even analytic) Jordan curves which have common points. We also have the problem of functions of class $L(k, \alpha)$ on Jordan arcs; Walsh [1935, p. 39] has shown that if f(z) is continuous on a Jordan arc then it can be uniformly approximated on that arc by polynomials. Theorem 3.2.2 is a result on degree of approximation in this connection but this is only the beginning. A circular arc is a special case which deserves attention; it is conceivable that a solution of this particular case might lead to more general results.

For lemniscates Curtiss [1941a] has obtained some highly interesting results on the relation between degree of convergence and continuity properties. In view of the fact that a critical point of a lemniscate (see §2.4) is a corner of two or more components of the lemniscate we thus have in the results of Curtiss [1941a] direct theorems corresponding to to Theorem 3.2.1 for curves with corners. Curtiss considers the Jacobi development and summations of various orders of this development and his results (in some cases the best possible) indicate the close relation between the Type t of the curve and the degree of convergence to be expected.

The constant M of Theorem 3.2.1 is of course independent of n and z but it depends upon the curve C; the type of this dependence (in particular upon the capacity of C) is an interesting question in itself and an investigation of it might well lead to important extensions of the above theorem. Also functions of class Log (k,1) merit attention with reference to degree of convergence of approximation sequences; we mention in this connection the more general problem of the relation between the modulus of continuity of a function and the degree of convergence of approximating polynomials.

The Faber polynomials are easily defined for non-analytic curves; their properties afford an interesting subject for investigation [compare Heuser, 1939, with references; Walsh and Sewell, 1940].

Also in connection with non-analytic curves a further study of the theorems of §3.4 should be profitable; examples for curves with corners ($t \neq 1$) would be interesting indeed. A complete investigation of the problems considered in §3.6 is an opportunity worthy of mention; under what conditions on the function and the boundary is it true that by integration we obtain a better degree of approximation, and how much better. Particular sequences might well be considered here; compare Theorems 6.2.3 and 6.2.4.

There are two further problems which are of particular interest and which, as far as we are aware, have not been mentioned in the literature. Suppose we modify the hypothesis of Theorem 3.4.1 by assuming inequality (3.4.1) valid not for every n but merely for a specific sequence of indices n_ν. What conditions on the sequence n_ν are sufficient to ensure the conclusion of Theorem 3.4.1? Under what conditions on the sequence n_ν can we derive a weaker conclusion than that of Theorem 3.4.1?

It has been pointed out that the Jackson summation depends upon k. An interesting problem in the real domain is the investigation of the possibilities of a summation with the desired properties and which is independent of k.

Of course the question of best approximation is with us always.

Chapter IV

APPROXIMATION MEASURED BY A LINE INTEGRAL

§4.1. DIRECT THEOREMS. The results of the preceding chapter on Tchebycheff approximation have immediate application to approximation in the sense of least weighted p-th powers. In fact the following theorem is a direct consequence of Theorem 3.2.1.

THEOREM 4.1.1. Let E, with boundary C, be a closed limited set bounded by a finite number of mutually exterior analytic Jordan curves, and let $f(z)$ belong to the class $L(k, \alpha)$ on C. Then there exist polynomials $P_n(z)$ such that

$$(4.1.1) \quad \int_C \Delta(z)|f(z)-P_n(z)|^p \, |dz| \leq M/n^{(k+\alpha)p}, \quad p > 0,$$

where $\Delta(z)$ is positive and continuous on C.

To show that this is in a sense the best possible degree of approximation we exhibit an example* used by Walsh and Sewell [1940] in connection with Problem β; in fact we return to this same example in appraising certain results of later chapters. Let C be the unit circle γ: $|z| = 1$ and let us set

---

* This example is closely akin to one used by Hardy and Littlewood [1932] in another connection.

$$(4.1.2)\quad f(z) = \sum_{\nu=0}^{\infty} \frac{z^{m_\nu}}{2^\nu},$$

where $m_{\nu+1}$ is a positive integer greater than m_ν ; this function is analytic in $|z| < 1$, and continuous in $|z| \leqq 1$. Let α be fixed, $0 < \alpha < 1$, and for ν sufficiently large let $m_{\nu+1}$ satisfy the inequality*

$$(4.1.3)\quad 2^{(\nu-1)/\alpha} < m_{\nu+1} - 1 \leqq 2^{\nu/\alpha} ;$$

the definition is consistent with the requirement $m_{\nu+1} > m_\nu$ for all ν , for the inequality $\nu+1 < 2^{(\nu-1)/\alpha}$ for ν sufficiently large follows from the inequality

$$\frac{\log(\nu+1)}{\nu - 1} < \frac{1}{\alpha} \log 2,$$

which is satisfied for ν sufficiently large. Moreover the relation

$$2^{\nu/\alpha} - 2^{(\nu-1)/\alpha} > 1$$

is satisfied for ν sufficiently large; it is merely the equivalent of the inequality

$$2^{\nu-1} > \frac{1}{(2^{1/\alpha} - 1)^\alpha} .$$

If we have $m_j \leqq n < m_{j+1}$ the partial sum $S_n(z)$ of degree n of the Taylor development of f(z) is

$$\sum_{\nu=0}^{j} \frac{z^{m_\nu}}{2^\nu} .$$

Thus for j sufficiently large we have on $|z| = 1$

$$|f(z) - S_n(z)| \leqq \frac{1}{2^j} \leqq \frac{1}{(m_{j+1} - 1)^\alpha} \leqq \frac{1}{n^\alpha},$$

---

* We remark that with the m_ν defined by (4.1.3) the function (4.1.2) cannot be continued beyond the unit circle; this is an immediate consequence of Hadamard's gap theorem [see, e.g., Dienes, 1931, pp. 231].

where $S_n(z)$ is the sum of the first n+1 terms of the Taylor development of f(z). Hence we conclude by Theorem 3.4.1 that f(z) belongs to the class $L(0, \alpha)$ on γ.

On the other hand, for $n = m_{j+1} - 1$ we have

$$(4.1.4) \quad \int_\gamma |f(z) - S_n(z)|^2 |dz| = \sum_{\nu=j+1}^{\infty} \frac{1}{2^{2\nu}}$$

$$= \frac{1}{2^{2j+2}}\left[1+\frac{1}{2^2} + \frac{1}{2^4} + \dots\right]$$

$$> \frac{1}{2^4 2^{2j-2}} = \frac{1}{2^4(2^{(j-1/\alpha)})^{2\alpha}}$$

$$> \frac{1}{2^4 n^{2\alpha}}.$$

But $S_n(z)$ is the polynomial of degree n of best approximation to f(z) in the sense of least squares (compare §4.3) and hence Theorem 4.1.1 is the best possible result for $k = 0$, $\alpha < 1$, in the sense that the exponent of n cannot be increased; in fact (4.1.4) shows that on the right hand side of inequality (4.1.1) the constant M cannot even be replaced by a function $\delta(n)$ where $\delta(n)$ approaches zero as n becomes infinite. The result extends to all positive k by considering successive indefinite integrals of f(z).

Neither can the exponent of n in inequality (4.1.1) be increased for an arbitrary function belonging to the class L(0,1). For suppose it could be shown that for every function of class L(0,1) the second member of inequality (4.1.1) could be replaced by $M_1/n^{2\beta}$, where β is some constant greater than unity. Then for the indefinite integral F(z) of the function f(z) above with $0 < \alpha < \beta - 1$, $\alpha < 1$,

$$F(z) = \sum_{\nu=0}^{\infty} \frac{z^{m_\nu +1}}{2^\nu (m_\nu + 1)},$$

which belongs to the class L(0,1), we have

$$\int_\gamma |F(z) - s_n(z)|^2 |dz| > \frac{M_2}{n^{(1+\alpha)2}},$$

where $s_n(z)$ is the polynomial of degree n of best approximation to F(z) in the sense of least squares. Thus for this particular function F(z) of class L(0,1) the second member of (4.1.1) cannot be replaced by $M_1/n^{2\beta}$ with $\beta > 1$. This discussion extends to all positive k. Hence Theorem 4.1.1 is the best possible in the sense that the exponent of n in the second member of (4.1.1) cannot be increased for $0 < \alpha \leq 1$, $k \geq 0$, $p = 2$.

For the line segment we have by virtue of Theorem 3.2.3

THEOREM 4.1.2. Let f(z) belong to the class $L(k, \alpha)$ on the segment C: $-1 \leq z \leq 1$. Then there exist polynomials $P_n(z)$ such that inequality (4.1.1) is valid.

§4.2. INDIRECT THEOREMS. We turn now to results in the converse direction which are far more difficult than the direct theorems of the preceding section. A given degree of approximation in the sense of least weighted p-th powers implies certain conditions on the function as is well known from classical results; however in view of the fact that we use the Lebesque integral it is clear that the value of the function on a point set of measure zero has no effect on the value of the left member of (4.1.1) and hence on the degree of approximation. For this reason our conclusions relate to the behavior of the function almost everywhere, rather than everywhere, on a set C of positive linear measure such as a curve or an arc. Of course if the function is originally assumed to be continuous then the conclusion defines the function uniquely throughout the set; for a rectifiable Jordan curve C the assumption that f(z) is continuous and can be approximated by polynomials on C leads to a definition of the function interior to C. In the theorems proved in the present section we assume the function defined merely on C and introduce a related function in the conclusion.

The method employed consists in establishing a Tchebycheff degree of approximation from the given degree of approximation in the sense of least weighted p-th powers; a powerful tool in this procedure is Theorem 2.2.1, which gives an inequality on the polynomial implied by a bound on the integral of the p-th power of the polynomial. Of course it is clear that once we have established a Tchebycheff degree of approximation to the function we have the results of §3.4 at our disposal. For obvious reasons it is sufficient to consider only a single curve or arc.

Let C be a rectifiable Jordan curve and let f(z) be defined on C, or almost everywhere on C. Also suppose that $P_n(z)$ exists such that

$$(4.2.1)\quad \int_C \Delta(z)|f(z)-P_n(z)|^p\,|dz| \leq \epsilon_n^p, \quad p > 0,$$

$$\epsilon_{n+1}^p \leq \epsilon_n^p, \quad \lim_{n\to\infty} \epsilon_n^p = 0,$$

where $\Delta(z)$ is positive and continuous on C. By virtue of the boundedness of $1/\Delta(z)$ we have

$$(4.2.2)\quad \int_C |f(z) - P_n(z)|^p\,|dz| \leq M_1\,\epsilon_n^p.$$

Let us write

$$(4.2.3)\quad f_1(z) = P_1(z) + [P_2(z) - P_1(z)] + \dots + [P_{n+1}(z) - P_n(z)] + \dots$$

and examine the behavior of this series. Since (4.2.2) holds for every n we have

$$(4.2.4)\quad \int_C |f(z) - P_{n+1}(z)|^p\,|dz| \leq M_1\,\epsilon_{n+1}^p;$$

hence the well known (see §2.5; Walsh [1935], p. 93) general inequalities

$$|X_1 + X_2|^p \leq 2^{p-1}|X_1|^p + 2^{p-1}|X_2|^p, \quad p > 1,$$

(4.2.5)

$$|x_1 + x_2|^p \leq |x_1|^p + |x_2|^p, \quad 0 < p \leq 1,$$

applied to (4.2.2) and (4.2.4) yield

$$\int_C |P_{n+1}(z) - P_n(z)|^p \, |dz| \leq M_2(\epsilon_n^p + \epsilon_{n+1}^p)$$
$$\leq M_3 \, \epsilon_n^p.$$

Thus it follows from Theorem 2.2.1 that

$$|P_{n+1}(z) - P_n(z)| \leq M_4 \, \epsilon_n (n+1)^{t/p}, \quad z \text{ on } \overline{C},$$

where C is a curve of Type t. Hence the series (4.2.3) converges uniformly and absolutely on C provided the series

(4.2.6) $$\sum_{\nu=1}^{\infty} \epsilon_\nu (\nu+1)^{t/p}$$

converges; in fact we have

$$|f_1(z) - P_n(z)| \leq M_4 \sum_{\nu=n}^{\infty} \epsilon_\nu (\nu+1)^{t/p}, \quad z \text{ on } \overline{C}.$$

Furthermore in case of the convergence of (4.2.6) the function $f_1(z)$ is analytic interior to C, continuous on $\overline{C}$, and the integral

(4.2.7) $$\int_C |f_1(z) - P_n(z)|^p \, |dz|$$

approaches zero as n becomes infinite. Applying the proper one of the inequalities (4.2.5) to (4.2.2) and (4.2.7) we see that

$$\int_C |f(z) - f_1(z)| \, |dz|$$

can be made arbitrarily small and hence, being a constant, it vanishes; thus $f(z) \equiv f_1(z)$ almost everywhere on C.

To be more specific let us suppose

$$\int_C \Delta(z)|f(z)-P_n(z)|^p \, |dz| \leq M/n^{\beta t}, \; p > 0, \; 1 \leq t \leq 2,$$

where $\Delta(z)$ is positive and continuous on C, and t is compatible with C. With n arbitrary but fixed it is advantageous to proceed as in the proof of Theorem 3.4.2 rather than use the series (4.2.3); we choose an integer m such that $2^{m-1} \leqq n < 2^m$ and write

$$f_1(z) - P_n(z) = [P_{2^m}(z) - P_n(z)] + P_{2^{m+1}}(z) - P_{2^m}(z)] + \ldots$$

Then by the same argument as above combined with the method used in the proof of Theorem 3.4.2 we have

$$|f_1(z)-P_n(z)| \leqq M_1/n^{(\beta-1)t/p}, \quad z \text{ on } \overline{C}, \quad \beta > 1.$$

We summarize the results in the following theorem.

THEOREM 4.2.1. Let C be a Jordan curve or arc of Type t and let f(z) be defined on C; let $P_n(z)$ exist such that

$$\int_C \Delta(z)|f(z)-P_n(z)|^p \, |dz| \leqq M/n^{\beta t}, \quad p > 0,$$

$$1 \leqq t \leqq 2, \quad \beta > 1,$$

where $\Delta(z)$ is positive and continuous on C. Then $f(z) = f_1(z)$ almost everywhere on C, where

$$f_1(z) = \lim_{n\to\infty} P_n(z), \quad z \text{ on } \overline{C};$$

Furthermore

$$|f_1(z)-P_n(z)| \leqq M/n^{(\beta-1)t/p}, \quad z \text{ on } \overline{C}.$$

By virtue of Theorem 3.4.1 we now have

THEOREM 4.2.2. Let C be a Jordan curve or arc of Type t and let f(z) be defined on C; let $P_n(z)$ exist such that

$$(4.2.8) \quad \int_C \Delta(z)|f(z)-P_n(z)|^p \, |dz| \leqq \frac{M}{n^{[(k+\alpha)p+1]t}}, \quad p > 0,$$

$$1 \leq t \leq 2, \; k > 0, \; 0 < \alpha \leq 1,$$

where $\Delta(z)$ is positive and continuous on C. Then $f(z) = f_1(z)$ almost everywhere on C, where

$$f_1(z) = \lim_{n \to \infty} P_n(z), \quad z \text{ on } \overline{C};$$

furthermore $f_1(z)$ belongs to the class $L(k, \alpha)$ on C if $\alpha < 1$, and to the class $Log(k, 1)$ on C if $\alpha = 1$.

As an example of an extremal function for C the unit circle γ: $|z| = 1$, $p = 2$, $\Delta(z) = 1$, we set

$$(4.2.9) \quad f(z) = (1-z)^{\alpha} = \sum_{\nu=0}^{\infty} a_{\nu} z^{\nu}, \quad 0 < \alpha < 1, \; |z| \leq 1,$$

where we choose the particular branch of the function which has the value unity at the origin. This function belongs to the class $L(0, \alpha)$ on γ but not to the class $L(0, \alpha')$, $\alpha' > \alpha$, on C. The fact that it does not belong to the class $L(0, \alpha')$ on γ follows by consideration of its behavior at the point $z = 1$. Since we need an evaluation of the coefficients here as well as in later applications we apply Theorem 3.4.1 to show that $f(z)$ belongs to the class $L(0, \alpha)$ on γ; also we carry through this evaluation for all real values of α other than the zero and the positive integers for future reference. In (4.2.9) we have

$$a_{\nu} = (-1)^{\nu} \frac{\alpha(\alpha-1) \ldots (\alpha - \nu + 1)}{\nu !}.$$

Let now α be positive and let c be the integral part of α, then

$$a_{\nu} = (-1)^{\nu} \frac{\alpha(\alpha-1) \ldots (\alpha-c)(\alpha-(c+1)) \ldots (\alpha-(\nu-1))}{\nu !}$$

$$= (-1)^{c+1} \alpha(\alpha-1) \ldots (\alpha-c) \left\{ \frac{(\nu-1-\alpha)(\nu-2-\alpha) \cdots (c+1-\alpha)}{\nu !} \right\}$$

$$= (-1)^{c+1} \alpha(\alpha-1) \ldots (\alpha-c) \frac{\Gamma(\nu-\alpha)}{\Gamma(c+1-\alpha)\Gamma(\nu+1)},$$

since $x\Gamma(x) = \Gamma(x+1)$, $x > 0$. By Stirling's formula

$$\Gamma(\nu) = \delta(\nu)\, \nu^{\nu-1/2} e^{-\nu},$$

$$\lim_{\nu\to\infty} \delta(\nu) = (2\pi)^{1/2}$$

we have

$$\frac{\Gamma(\nu-\alpha)}{\Gamma(\nu+1)} = \nu^{-\alpha-1}(1+\epsilon_\nu), \quad \lim_{\nu\to\infty} \epsilon_\nu = 0;$$

hence it follows that

$$a_\nu = \frac{\Gamma(\alpha)}{\Gamma(c+1-\alpha)} \nu^{-\alpha-1}(1+\epsilon_\nu).$$

For α negative by setting $|\alpha| = \beta$ we have

$$a_\nu = \frac{\nu^{\beta-1}}{\Gamma(\beta)}(1+\epsilon_\nu).$$

Now since a_ν is asymptotic to $\nu^{-\alpha-1}$ it is clear that for $0 < \alpha < 1$ we have

$$\left| f(z) - \sum_{\nu=0}^{n} a_\nu z^\nu \right| \leq \sum_{n+1}^{\infty} |a_\nu| \leq M/n^\alpha, \quad |z| \leq 1,$$

and hence by Theorem 3.4.1 the function $f(z)$ belongs to the class $L(0, \alpha)$ on γ; as mentioned above we have merely to set $z_1 = 1$ and $z_2 = z$ to see that $f(z)$ satisfies a Lipschitz condition of no order higher than α on γ. Thus the following inequalities

$$\begin{aligned} \int_\gamma \left| f(z) - \sum_{\nu=0}^{n} a_\nu z^\nu \right|^2 |dz| &= 2\pi \sum_{n+1}^{\infty} |a_\nu|^2 \\ &\leq M_1 \sum_{n+1}^{\infty} \frac{1}{\nu^{2\alpha+2}} \\ &\leq M_2 / n^{2\alpha+1} \end{aligned}$$

lead us to the conclusion that Theorem 4.2.1 cannot be improved as far as concerns the relation between the exponent of n and the class of $f_1(z)$, at least for the unit circle, $k = 0$, $\alpha < 1$, $p = 2$. For $\alpha = 1$ we use the same argument as for the corresponding situation in §4.1; of course the entire discussion extends to arbitrary positive integral k by integrating a suitable number of times.

For rectifiable Jordan arcs the value of t may be 2 and there is no corresponding region of analyticity.

In connection with approximation measured by a line integral Hardy and Littlewood [1928] have stated without proof a beautiful result. A function $f(\theta)$ which is periodic of period 2π is said to satisfy an <u>integrated Lipschitz condition of order α with exponent p</u>, $p > 0$, provided

$$(4.2.10)\quad \left[\frac{1}{2\pi}\int_{-\pi}^{\pi}|f(\theta+h) - f(\theta)|^p\, d\theta\right]^{1/p} \leq Lh^{\alpha},$$

where L is a constant independent of h. According to Hardy and Littlewood if $f(\theta)$ satisfies condition (4.2.10) there exist trigonometric sums $S_n(\theta)$ of order n such that we have

$$(4.2.11)\quad \left[\frac{1}{2\pi}\int_{-\pi}^{\pi}|f(\theta) - S_n(\theta)|^p\, d\theta\right]^{1/p} \leq M/n^{\alpha},$$

and conversely if there exist $S_n(\theta)$ such that inequality (4.2.11) is valid then the function $f(\theta)$ satisfies condition (4.2.10). In other words the class of functions satisfying an integrated Lipschitz condition of order α with exponent p is identical with the class of functions which can be approximated in the mean with exponent p by trigonometric sums of order n with error not greater than M/n^{α}.

§4.3. ORTHOGONAL POLYNOMIALS. We resume the study of the preceding section where we now take $p = 2$ and $\Delta(z) \equiv 1$. We are thus studying approximation in the sense of least squares with weight function unity, which is the most important case of approximation in the sense of least weighted p-th powers.

Let C be a rectifiable Jordan arc or curve; the set of polynomials $Q_n(z)$, $n = 0,1,2,\ldots$, is said to be orthogonal on C if we always have

$$\int_C Q_m(z)\overline{Q}_n(z)\ |dz| = 0, \quad m \neq n,$$

and is said to be normal on C if we have

$$\int_C Q_n(z)\overline{Q}_n(z)\ |dz| = 1.$$

The well known Gram-Schmidt method of orthogonalization and normalization applied to the sequence of polynomials $1, z, z^2, \ldots$ on the curve or arc C yields a set of polynomials $Q_0(z)$, $Q_1(z)$, $Q_2(z), \ldots$, normal and orthogonal on C.

Let f(z) belong to the class $L(k, \alpha)$ on C, then the formal development of f(z) in the set of Polynomials $Q_n(z)$, normal and orthogonal on C, is

$$(4.3.1) \quad f(z) \backsim a_0Q_0(z) + a_1Q_1(z) + \ldots + a_\nu Q_\nu(z) + \ldots,$$

$$a_\nu = \int_C f(z)\ \overline{Q}_\nu(z)\ |dz|;$$

here the sign $\backsim$ is used simply to denote formal correspondence. Let

$$(4.3.2) \quad S_n(z) = a_0Q_0(z) + a_1Q_1(z) + \ldots + a_nQ_n(z);$$

this polynomial $S_n(z)$ is the polynomial of degree n of best approximation to f(z) on C in the sense of least squares. Other properties of this expansion are important, such as the following

$$(4.3.3) \quad \begin{aligned} \int_C |f(z) - S_n(z)|^2 |dz| &= \sum_{\nu=n+1}^{\infty} |a_\nu|^2, \\ \int_C |f(z)|^2 |dz| &= \sum_{\nu=0}^{\infty} |a_\nu|^2; \end{aligned}$$

of course these series converge since $|f(z)|^2$ is integrable on C. It is clear from (4.3.3) that degree of convergence can be studied through the coefficients of the orthogonal development. Various other important properties are well known [see, e.g., Walsh 1935; Szegö 1939; Smirnoff 1928, 1932].

Suppose C consists of a finite number of mutually exterior analytic Jordan curves and suppose $f(z)$ belongs to the class $L(k, \alpha)$ on C; then by Theorem 4.1.1 and the least square property of $S_n(z)$ we know that

$$\int_C |f(z) - S_n(z)|^2 |dz| \leq M/n^{(k+\alpha)\cdot 2},$$

or by (4.3.3)

$$\sum_{\nu=n+1}^{\infty} |a_\nu|^2 \leq M/n^{(k+\alpha)\cdot 2}.$$

Consequently we obtain

$$|a_n| \leq M_1/n^{k+\alpha},$$

and the theorem

THEOREM 4.3.1. Let C consist of a finite number of mutually exterior analytic Jordan curves and let $f(z)$ belong to the class $L(k, \alpha)$ on C; let $Q_n(z)$ be the set of polynomials normal and orthogonal on C and let

$$f(z) = \sum_{\nu=0}^{\infty} a_\nu Q_\nu(z), \quad a_\nu = \int_C f(z)\overline{Q}_\nu(z)|dz|.$$

Then

$$|a_n| \leq M/n^{k+\alpha}.$$

We have a similar result for the line segment; the statement and proof of this theorem is left to the reader.

It is easy to see that in the converse direction inequalities on the coefficients yield results on degree of approximation in the sense of least squares; for convenience in studying the details in this direction we need the following definition.

DEFINITION 4.3.2. Let C be a rectifiable

Jordan curve and let $Q_n(z)$ be the set of polynomials normal and orthogonal on C; let f(z) be analytic interior to C and continuous in $\overline{C}$. We say that f(z) is of class F_q on C if the numbers

$$(4.3.4) \quad n^q \int_C f(z)\overline{Q}_n(z)|dz|, \ q > 0,$$

are uniformly bounded.

From this definition and equations (4.3.3) we have

THEOREM 4.3.3. If f(z) is of class F_q on C then

$$(4.3.5) \quad \int_C |f(z) - S_n(z)|^2 |dz| \leq M/n^{2q-1}, q > \tfrac{1}{2}.$$

A comparison of Theorems 4.1.1 and 4.3.3 shows that we establish the same inequalities for a function belonging to the class $L(0,q-\frac{1}{2})$ on C, assuming $0 < q-\frac{1}{2} \leq 1$ (the extension to derivatives is immediate), as for a function of class F_q on C. It is not to be concluded, however, that these two classes are identical. There is a connection (not reciprocal) which is easy to establish in the case of an analytic Jordan curve. From Theorem 4.3.1 and Definition 4.3.2 it follows that a function of class $L(k, \alpha)$ on C is also of class $F_{k+\alpha}$ on C. In the other direction let $q = k + \alpha$, $0 < \alpha < 1$, $k \geq 1$, and let f(z) belong to the class F_q on C; then by applying the conclusion of Theorem 4.2.1 to Theorem 4.3.3 we have

$$|f(z) - S_n(z)| \leq M_1/n^{q-1} = M_1/n^{k-1+\alpha}, \ z \text{ on } \overline{C}.$$

Now by Theorem 3.4.1 it follows that f(z) belongs to the class $L(k-1, \alpha)$ on C. Our theorem is

THEOREM 4.3.4. Let C be an analytic Jordan curve and let f(z) be of class $F_{k+\alpha+1}$, $0 < \alpha < 1$, on C. Then $f(z)$ belongs to the class $L(k, \alpha)$ on C.

In connection with Theorem 4.3.4 let us consider the function $f(z) = (1-z)^{\alpha}$, $0 < \alpha < 1$, $f(0) = 1$. For the unit circle γ: $|z| = 1$ the polynomials 2π, $(2\pi)^{1/2}z$, $(2\pi)^{1/2}z^2$, ..., $(2\pi)^{1/2}z^n$, are normal and orthogonal on γ ($\Delta(z) \equiv 1$) and hence the Taylor development is a development of the type we are considering; thus it is clear from the asymptotic property of the Taylor coefficients (see §4.2) that $f(z)$ is of class $F_{1+\alpha}$ on γ, and yet we have already seen (also §4.2) that $f(z)$ satisfies a Lipschitz condition of no order $\beta > \alpha$.

For C an analytic Jordan curve Szegö [1939] has established the asymptotic formula

$$(4.3.6) \qquad Q_n(z) \sim \frac{1}{(2\pi)^{1/2}} [\phi'(z)]^{1/2}[\phi(z)]^n/w(z),$$

where $\phi(z)$ is the usual mapping function, $w(z)$ is analytic and different from zero in K even at infinity, and $|w(z)|^2$ is continuous in K and equal to $\Delta(z)$ on C. The asymptotic relation is valid on any set exterior to C in the sense that the ratio of the two members approaches unity uniformly as n becomes infinite. If $\Delta(z)$ is analytic in the closed exterior of C then (4.3.6) is valid on C and even in a sufficiently small neighborhood interior to C. We continue to assume $\Delta(z) \equiv 1$; then by virtue of the uniform boundedness of $|Q_n(z)|$ on C we have

THEOREM 4.3.5. Let C be an analytic Jordan curve and let $f(z)$ belong to the class F_q on C, $q > 1$. Then with the notation (4.3.2) we have

$$|f(z) - S_n(z)| \leq M/n^{q-1}, \quad z \text{ on } \overline{C}.$$

Of course it likewise follows from the boundedness of $|Q_n(z)|$ that if $\sum_{\nu=0}^{\infty} |a_\nu|$ converges the function

$f(z)$ is continuous on C.

The theorems and definitions of this section are due to Walsh and Sewell [1940].

§4.4. POLYNOMIALS OF BEST APPROXIMATION. In §4.3 we studied the polynomials of best approximation in the sense of least squares; in the present section we consider the general case, that is, polynomials of best approximation in the sense of least weighted p-th powers, $p > 0$. Let C be a rectifiable Jordan arc or curve and let $f(z)$ belong to the class $L(k, \alpha)$ on C; let the weight function $\Delta(z)$ be positive and continuous on C. Under these conditions it is known* that there exists a polynomial $p_n(z)$ of best approximation to $f(z)$ in the sense of least weighted p-th powers; if $p > 1$ the polynomial $p_n(z)$ is unique.

If we set

$$(4.4.1) \qquad \int_C \Delta(z)\left|f(z) - p_n(z)\right|^p |dz| = \epsilon_n, \quad p > 0,$$

then from the preceding theorems we can write down certain conditions on ϵ_n. For example if C is analytic and if $f(z)$ belongs to the class $L(k, \alpha)$ we know from Theorem 4.1.1 that $\epsilon_n n^{(k+\alpha)p}$ is bounded; we also know from the example of §4.1.1 that given $\alpha' > \alpha$ there exist functions of class $L(k, \alpha)$ for which the quantity $\epsilon_n n^{(k+\alpha')p}$ is not bounded.

With regard to the Tchebycheff degree of approximation of the polynomials $p_n(z)$ we prove the following theorem:

THEOREM 4.4.1. Let C be a Jordan curve of Type t; let $f(z)$ be analytic interior to C and continuous on $\overline{C}$; let $p_n(z)$ be a polynomial which minimizes (4.4.1), where $\Delta(z)$ is positive and continuous on C. Let $P_n(z)$ exist such that

* See, e.g., Walsh [1935], Chap. XII.

$$(4.4.2)\quad |f(z) - P_n(z)| \leq \delta_n, \quad z \text{ on } \overline{C}.$$

Then we have

$$(4.4.3)\quad |f(z) - p_n(z)| \leq Mn^{t/p}\,\delta_n,$$
$$z \text{ on } \overline{C},\ 1 \leq t \leq 2.$$

The proof is similar to that of Theorem 2.2.1. In the identity $f(z)-p_n(z) \equiv f(z) - P_n(z) + P_n(z)-p_n(z)$ we consider the polynomial $\Pi_n(z) = P_n(z)-p_n(z)$. Suppose $|\Pi_n(z)| \leq \mu_n$, z on C, and $|\Pi_n(z_0)| = \mu_n$, z_0 on C; then by the argument and in the notation of §2.2 we have

$$(4.4.4)\quad \epsilon_n \equiv \int \Delta(z)|f(z) - p_n(z)|^p\,|dz| \geq$$
$$\geq M_0\left(\frac{2A-2}{2A}\right)^p \mu_n^p \quad \frac{1}{AK(C)n^t},$$

provided $\mu_n \geq \delta_n/2A$, where M_0 depends on $\Delta(z)$; the minimizing property of $p_n(z)$ yields

$$(4.4.5)\quad \epsilon_n \leq M_1\,\delta_n^p,$$

since $\Delta(z)$ is positive and continuous. Combining inequalities (4.4.4) and (4.4.5) we have

$$\mu_n \leq M_2\, n^{t/p}\quad \delta_n;$$

thus whether $\mu \geq \delta_n/2A$ or not we have (4.4.3) and the proof is complete.

In considering the relation between best approximation in the sense of Tchebycheff and in the sense of least weighted p-th powers it is interesting to note that for each n we have*

$$\lim_{p\to\infty} p_n(z) = T_n(z),$$

where $p_n(z)$ is the polynomial of degree n of best approximation to f(z) in the sense of least weighted p-th powers

---

* Julia [1926]; the result for C the segment (0,1) of the axis of reals is due to Pólya [1915].

and $T_n(z)$ is the polynomial of best approximation to $f(z)$ in the sense of Tchebycheff. We see by Theorem 4.4.1 that the Tchebycheff degree of approximation of the polynomials of best approximation in the sense of least weighted p-th powers approaches the degree of best approximation in the sense of Tchebycheff as p becomes infinite; this remark is to be understood to mean that the right hand member of (4.4.3) approaches the second member of (4.4.2) except for the constant factor M as p becomes infinite.

It is clear from the method of proof that Theorem 4.4.1 is valid also if C is a rectifiable Jordan arc; in this case $t = 2$.

Using the results of Theorem 3.2.1 to evaluate δ_n in Theorem 4.4.1 we have

COROLLARY 4.4.2. In Theorem 4.4.1 let C be an analytic Jordan curve and let $f(z)$ belong to the class $L(k, \alpha)$ on C. Then we have

$$|f(z) - p_n(z)| \leq \frac{M}{n^{k+\alpha-1/p}}, \quad z \text{ on } \overline{C}.$$

For the orthogonal polynomials $p = 2$; thus we have in the notation of §4.3

$$|f(z) - S_n(z)| \leq \frac{M}{n^{k+\alpha-1/2}}, \quad z \text{ on } \overline{C}.$$

Theorem 4.4.1 was proved by Jackson [1930a] with $t = 1$ for C a Jordan curve whose parametric representation possessed a second derivative bounded from zero; he [1931] also proved the theorem with $t = 2$ under lighter restrictions on C. The theorem as stated here as well as the corollary is due to Sewell [1937a]; the method of proof is that used by Jackson.

§4.5. GENERALITY OF THE WEIGHT FUNCTION. Throughout our discussion of approximation in the sense of least weighted p-th powers we have assumed the weight function $\Delta(z)$ positive and continuous on each curve or arc. This restriction can be lightened, however, without altering the conclusions of the theorems in general. Of course the asymptotic formula (4.3.6) for orthogonal polynomials has been shown to be valid on C only if C is analytic and if $\Delta(z)$ is analytic; thus it is clear that in Theorem 4.3.5 the analyticity of $\Delta(z)$ must be included in the hypothesis.

In other situations, however, an analysis of the method leads to lighter conditions. In one direction we need to obtain an inequality of the form $\int_C |P_n(z)|^p \, |dz| \leq ML^p$ from an inequality such as $\int_C \Delta(z) |P_n(z)|^p \, |dz| \leq L^p$; in fact we have

$$\int_C \Delta(z) \, |P_n(z)|^p \, |dz| \geq \left[\underset{z \text{ on } C}{\text{g.l.b.}} \Delta(z)\right] \int_C |P_n(z)|^p \, |dz|,$$

provided $\Delta(z)$ is positive, bounded from zero, and integrable. This applies to the results of §§4.2 and 4.3.

If the nonnegative integrable norm function $\Delta(z)$ is not bounded from zero but some negative power of $\Delta(z)$ is integrable we obtain a somewhat weaker result by means of the Hölder inequality (which for $\sigma = \frac{1}{2}$ becomes the Schwarz inequality)

$$(4.5.1) \qquad \left|\int F^{\sigma} G^{1-\sigma}\right| \leq \left(\int |F|\right)^{\sigma} \left(\int |G|\right)^{1-\sigma}, \quad 0 < \sigma < 1,$$

whose validity requires merely the existence of the integrals on the right. Suppose for example $[\Delta(z)]^{-\beta}$, $\beta > 0$, is integrable, then we can set $\sigma = 1/(1+\beta)$:

$$\int |P_n(z)^{p(1-\sigma)}| dz|$$

$$\leq \left[\int \frac{|dz|}{[\Delta(z)]^{(1-\sigma)/\sigma}}\right]^{\sigma} \left[\int \Delta(z) |P_n(z)|^p |dz|\right]^{1-\sigma}$$

$$\leq M_1 \, (L^p)^{1-\sigma} = M_1 \, L^{p(1-\sigma)};$$

thus by the method of proof of Theorem 2.2.1 the inequality

$$\int_C \Delta(z)|P_n(z)|^p \, |dz| \leq L^p, \; p > 0,$$

yields the inequality

$$|P_n(z)| \leq L' L n^{t/[p(1-\delta)]}, \quad \delta < 1, \; z \text{ on } C.$$

Thus if we modify the hypothesis of Theorem 4.2.2 by requiring that the weight function $\Delta(z)$ no longer be positive and continuous but merely that it be nonnegative and integrable and that $[\Delta(z)]^{-\beta}$, $\beta > 0$, be integrable, and if we replace the right member of inequality (4.2.8) by the quantity

$$\frac{M}{n^{[(k+\alpha)p + 1/(1-\delta)]t}},$$

where $\delta = 1/(\beta+1)$ the conclusion of Theorem 4.2.2 remains valid.

In the other direction from an inequality of the form $|f(z) - P_n(z)| \leq \epsilon_n$, z on C, we arrived at an inequality of the form

$$\int_C \Delta(z)|f(z) - P_n(z)|^p \, |dz| \leq M(\epsilon_n)^p ;$$

as a matter of fact all that is required here is that $\Delta(z)$ be non-negative and integrable in the sense of Lebesgue, for

$$\int_C \Delta(z)|f(z)-P_n(z)|^p \, |dz| \leq \underset{z \text{ on } C}{\text{l.u.b.}} |f(z)-P_n(z)|^p \int_C \Delta(z)|dz|.$$

Such generalizations have been pointed out by Jackson [1930] and Walsh [1935].

§4.6. EXERCISES. 4.6.1. Let C be an analytic Jordan curve; then in the notation of §3.3 we have

$$\int_C \Delta(z)\left|f(z) - \sum_{\nu=0}^{n} d_{n,\nu} \, a_\nu \, F_\nu(z)\right|^p |dz| \leq \frac{M}{n^{(k+\alpha)p}}, \quad p > 0,$$

NOTE. Throughout these exercises we assume $\Delta(z)$ positive and continuous on C.

4.6.2. Under the hypothesis of Ex. 4.6.1 we have

$$\int \Delta(z) \left| f(z) - \sum_{\nu=0}^{n} a_\nu F_\nu(z) \right|^p |dz| \leq M \left[\frac{\log n}{n^{k+\alpha}} \right]^p, \quad p > 0.$$

4.6.3. Let $f(z) = \sum_{\nu=0}^{\infty} a_\nu z^\nu$ be analytic in $|z| < 1$ and continuous in $|z| \leq 1$; let $f'(z)$ be continuous on $|z| \leq 1$. Then

$$\int_{|z|=1} \left| f'(z) - \sum_{\nu=0}^{n} \nu a_\nu z^{\nu-1} \right|^2 |dz| \geq n^2 \int_{|z|=1} \left| f(z) - \sum_{\nu=0}^{n} a_\nu z^\nu \right|^2 |dz|.$$

[Walsh and Sewell, 1940]

4.6.4. Let C be a Jordan curve of Type t, then we have $|Q_n(z)| \leq Mn^{t/2}$, z on $\overline{C}$. [Sewell, 1937a]

NOTE. In Ex. 4.6.4 and hereafter we denote by $Q_n(z)$ the polynomials normal and orthogonal on C with $\Delta(z)$ as specified above.

4.6.5. In Ex. 4.6.4 if f(z) is of class F_q on C then (in the notation of §§4.2 and 4.3) $f(z) = f_1(z)$ almost everywhere on C, and

$$|f_1(z) - S_n(z)| \leq \frac{M}{n^{(2q-t-1)/2}}, \quad z \text{ on } \overline{C}.$$

4.6.6. If in Ex. 4.6.5 we have $(2q-t-1)/2t = k + \alpha$, then $f_1(z)$ belongs to $L(k, \alpha)$ on C if $0 < \alpha < 1$, and to Log (k,1) on C if $\alpha = 1$.

4.6.7. Let C be a rectifiable Jordan curve and let f(z) be continuous on C; let $\sum_{\nu=0}^{\infty} |a_\nu| \nu^{t/2}$ converge. Then $f(z) = f_1(z)$ where $f_1(z) = \sum_{\nu=0}^{\infty} a_\nu Q_\nu(z)$ is analytic interior to C, and continuous in $\overline{C}$. [Sewell, 1937a]

4.6.8. Let $f(z) = \sum_{\nu=0}^{\infty} a_\nu z^\nu$ belong to the class $L(k, \alpha)$ on $\gamma: |z| = 1$. Then

$$\iint_{\overline{\gamma}} \left| f(z) - \sum_{\nu=0}^{n} a_\nu z^\nu \right|^2 dS \leq \frac{M}{n^{2(k+\alpha)+1}}.$$

NOTE. Use the fact that $1, z, z^2, \ldots,$ are orthogonal on $|z| < 1$; compare Walsh [1935, p. 149].

4.6.9. Let C be a piecewise analytic Jordan curve

of Type t, $1 \leq t < 2$; let $R_n(z)$ be normal and orthogonal on $\overline{C}$. Then $|R_n(z)| \leq Mn^t$, z on $\overline{C}$.

NOTE. Here and hereafter we denote by $R_n(z)$ the polynomials normal and orthogonal on the surface $\overline{C}$; the definition is entirely similar to that of $Q_n(z)$ except that the line integral is replaced by the surface integral; $b_n = \iint_{\overline{C}} \Delta(z) f(z) \overline{R}_n(z)\, dS$.

4.6.10. For the polynomials $R_n(z)$ normal and orthogonal on $|z| \leq 1$ we have $R_n(z) = (n+1)^{1/2} z^n/(\pi)^{1/2}$.

4.6.11. Let C be an analytic Jordan curve and let $f(z)$ belong to the class $L(k, \alpha)$ on C. Then

$$|b_n| \leq M/n^{k+\alpha},$$

in the notation of Ex. 4.6.9.

DEFINITION. A function $f(z)$ is of class F_q on $\overline{C}$, where C is a Jordan curve, if the numbers $n^q b_n$ are uniformly bounded.

4.6.12. Let C be a Jordan curve and let $f(z)$ be of class F_q on $\overline{C}$. Then we have

$$\iint_{\overline{C}} \left| f(z) - \sum_{\nu=0}^{n} b_\nu R_\nu(z) \right|^2 dS \leq M/n^{2q-1}, \quad q > 1/2$$

4.6.13. Let E, with boundary C, be a closed limited set in the z-plane bounded by the mutually exterior analytic Jordan curves $C_1, C_2, \ldots, C_\lambda$. Let $f(z)$ belong to the class $L(k, \alpha)$ on C. Then $P_n(z)$ exists such that

$$\iint_E \Delta(z) \left| f(z) - P_n(z) \right|^p dS \leq M/n^{(k+\alpha)p}, \quad p > 0,$$

where $\Delta(z)$ is positive and continuous on each C_j.

4.6.14. Let C be a Jordan curve composed of a finite number of analytic arcs meeting in corners of exterior openings $\mu_j \pi$, $1 \leq \mu_j < 2$ and let $f(z)$ be defined on $\overline{C}$. Let $P_n(z)$ exist such that

$$\iint_{\overline{C}} \Delta(z) \left| f(z) - P_n(z) \right| dS \leq M/n^{\beta t}, \quad \beta > 2,$$

where t is the Type of C. Then $f(z) = f_1(z)$ almost everywhere on $\overline{C}$, where $f_1(z)$ is analytic in C, continuous in $\overline{C}$, and

$$|f_1(z) - P_n(z)| \leq M/n^{(\beta-2)t/p}, \quad z \text{ on } \overline{C}.$$

SUGGESTION: Use the method of Theorem 4.2.1 along with Ex. 2.4.19.

4.6.15. In Ex. 4.6.14 if $(\beta-2)/p = k+\alpha$, then $f_1(z)$ belongs to the class $L(k,\alpha)$ on C if $0 < \alpha < 1$, and to Log $(k,1)$ on C if $\alpha = 1$.

4.6.16. Let C be the curve of Ex. 4.6.14, and let $f(z)$ be analytic interior to C and continuous in $\overline{C}$. Let $p_n(z)$ be a polynomial which minimizes

$$\iint_{\overline{C}} \Delta(z)|f(z) - p_n(z)|^p \, dS, \quad p > 0,$$

where $\Delta(z)$ is positive and continuous on $\overline{C}$; let $P_n(z)$ be a polynomial such that

$$|f(z) - P_n(z)| \leq \delta_n, \quad z \text{ on } \overline{C}.$$

Then we have

$$|f(z) - p_n(z)| \leq Mn^{2t/p}\,\delta_n, \quad z \text{ on } \overline{C},$$

where t is the type of C.

4.6.17. In Ex. 4.6.16 let C be analytic and let $F(z)$ belong to the class $L(k, \alpha)$ on C. Then

$$|f(z) - p_n(z)| \leq M/n^{k+\alpha-2/p}, \quad z \text{ on } \overline{C}.$$

4.6.18. Let C be the unit circle. Let $f(z) = \sum_{\nu=0}^{\infty} a_\nu z^\nu$ be analytic in $|z| < 1$ and let $f'(z)$ be continuous in $|z| \leq 1$. If we have

$$\int_C \left| f(z) - \sum_{\nu=0}^{n} a_\nu z^\nu \right|^2 |dz| = \epsilon_n^2,$$

then we have

$$\int_C \left| f'(z) - \sum_{\nu=1}^{n} \nu a_\nu z^{\nu-1} \right|^2 |dz| \geq Mn^2 \epsilon_n^2.$$

[Walsh]

4.6.19. In the notation of Ex. 4.6.18 give simple conditions on the coefficients a_ν so that

$$\int_C \left| f'(z) - \sum_{\nu=1}^{n} \nu a_\nu z^{\nu-1} \right|^2 |dz| \leq Mn^2 \epsilon_n^2 .$$

[Walsh]

§4.7. DISCUSSION. Our results are fairly complete as far as analytic Jordan curves are concerned, but there is the same problem here as in Chapter III of extending such results as Theorem 4.1.1 to more general curves. Also there accompanies this problem the question of examples; for instance an example appraising Theorem 4.1.2 would be an important contribution.

In Ex. 4.6.4 we have an upper bound on $Q_n(z)$; it is not known whether this is the best possible in general. In fact very few results are available on the properties of $Q_n(z)$ for other than analytic Jordan curves. Also the concept of classes L_2 and H_2 should be extended to arbitrary rectifiable Jordan curves and the corresponding theory of approximation studied [see Smirnoff, 1928, 1932; Kildysch and Lavrentieff, 1937].

There are many interesting unsolved problems connected with polynomials orthogonal on several curves [compare Walsh, 1935a, p. 43]. Let the set of polynomials $Q_\nu(z), \nu = 0, 1, 2, \ldots$, be orthogonal on the curve C_1 with respect to the weight function $\Delta_1(z)$. When is this same set of polynomials orthogonal on another curve C_2 with respect to a weight function $\Delta_2(z)$? Given C_1 can $\Delta_1(z)$ always be found so that C_2 and $\Delta_2(z)$ exist? Given C_1 and C_2, can $\Delta_1(z)$ and $\Delta_2(z)$ always be found? These problems have been studied by various authors: Walsh [1934, 1935a]; Geronimus [1931, 1940]; Szegö [1935, 1939a]; Merriman [1938]. There are also the corresponding problems concerning harmonic polynomials; compare Walsh and Merriman [1937].

It is also of interest in this connection to study orthogonality not merely as defined in §4.3 but also in the sense that

$$\int_C \Delta(z)Q_m(z)Q_n(z)dz = 0;$$

compare Walsh [1935a], Walsh and Merriman [1937], Geronimus [1931, 1940].

In all of these problems the situation concerning lemniscates deserves attention.

One has merely to glance at the exercises on surface integral approximation in the sense of least weighted p-th powers to see that a systematic study of this theory has not yet been undertaken as far as Problem α is concerned. Important contributions to this field are due to Bergman [1922], Bochner [1922], and Carleman [1922]; an extension of Carleman's asymptotic formula for $R_n(z)$ in K to the curve C itself would yield new results on approximation. A comparison of Ex. 4.6.8 with Ex. 4.6.13 raises an important question in this theory. It should be noted in this connection that in Ex. 4.6.8 the polynomial is interpolating to $f(z)$ in $n + 1$ points coincident in the origin and thus the difference may be small throughout a large area of the surface; it is quite conceivable that there may be polynomials converging to $f(z)$ with the best degree of approximation in the sense of Tchebycheff for which we have only the degree concluded in Ex. 4.6.13. As in Ex. 2.4.19 the case $t = 2$ should be studied in connection with Exs. 4.6.9, 4.6.14, 4.6.15, and 4.6.16.

PART II. PROBLEM β

CHAPTER V

PRELIMINARIES

§5.1. AN INTERPOLATION FORMULA AND INEQUALITIES. The classical interpolation formula (2.3.1) or (2.3.2) is fundamental in the study of the relation between degree of approximation and regions of analyticity; however, this formula is not well adapted to a study of Problem β due to the fact that the properties of $f(z)$ on C_ρ cannot be effectively utilized to reflect in the degree of approximation to $f(z)$ on C or on E. In our study of Problem β the following modification is highly important in the light of results already established on Problem α and the evaluations of §2.4 and §2.5.

THEOREM 5.1.1. Let E, with boundary C, be a closed limited set whose complement K is connected and regular. Let $f(z)$ be analytic interior to C_ρ and representable by the Cauchy integral over C_ρ. Let the points $z_1, z_2, \ldots, z_{n+1}$ lie interior to C_ρ, and let

$$(5.1.1)\quad \omega_n(z) \equiv (z-z_1)(z-z_2)\ldots(z-z_{n+1}).$$

Then we have

$$(5.1.2)\quad f(z) - p_n(z)$$
$$= \frac{1}{2\pi i}\int_{C_\rho}\frac{\omega_n(z)\cdot[f(t)-P_n(t)]dt}{\omega_n(t)\cdot(t-z)},\ z \text{ interior to } C_\rho,$$

where $P_n(z)$ is an arbitrary polynomial, and where $p_n(z)$ is the unique polynomial interpolating to $f(z)$ in the points z_j, $j = 1,2,\ldots, n+1$.

Under the hypothesis of the theorem we have the formula

$$(2.3.2)\quad f(z) - p_n(z) \equiv \frac{1}{2\pi i}\int_{C_\rho} \frac{\omega_n(z)f(t)dt}{\omega_n(t)(t-z)},$$

z interior to C_ρ.

In particular we may set $f(z) \equiv P_n(z)$ in (2.3.2), then the polynomials $p_n(z)$ and $P_n(z)$ coincide in $n + 1$ points z_j and hence are identically equal. Thus we have

$$(5.1.3)\qquad 0 \equiv \frac{1}{2\pi i}\int_{C_\rho} \frac{\omega_n(z)P_n(t)}{\omega_n(t)(t-z)}\,dt,\ z \text{ interior to } C_\rho .$$

If we subtract (5.1.3) from (2.3.2) in its original form (i.e., without particularizing $f(z)$), we have (5.1.2) and the proof of the theorem is complete.

It is clear that in (5.1.2) under certain conditions on $f(z)$ we can use the results concerning the degree of convergence of a sequence $P_n(z)$ to $f(z)$ on C_ρ (Chap.III) and the inequalities of §2.5 to obtain theorems on the degree of convergence of the sequence $p_n(z)$ to $f(z)$ on E (Problem β). For instance by inspection of (5.1.2) we have the useful inequality

$$(5.1.4)\quad |f(z) - p_n(z)| \leq \frac{1}{2\pi}\int_{C_\rho} \left|\frac{\omega_n(z)}{\omega_n(t)}\right| \frac{|f(t)-P_n(t)|\ |dt|}{|t-z|},$$

z interior to C_ρ.

Thus we have

THEOREM 5.1.2. In Theorem 5.1.1 suppose

$$\left|\frac{\omega_n(z)}{\omega_n(t)}\right| \leq M/\rho^n,\ z \text{ on } C,\ t \text{ on } C_\rho ,$$

$$\left| f(z) - P_n(z) \right| \leqq \epsilon_n, \; z \text{ on } C_\rho \; .$$

Then we have

$$\left| f(z) - p_n(z) \right| \leqq M_1 \, \epsilon_n / \rho^n, \; z \text{ on } \overline{C}.$$

Thus through a Tchebycheff degree of approximation to $f(z)$ on C_ρ we obtain a Tchebycheff degree of approximation to $f(z)$ on C.

By means of the Hölder inequality (4.5.1), we have

$$(5.1.5) \quad \left| f(z) - p_n(z) \right|$$

$$\leqq \frac{1}{2\pi} \left[\int_{C_\rho} \left| \frac{\omega_n(z)}{\omega_n(t)(t-z)} \right|^{1/\delta} |dt| \right]^{\delta} \left[\int_{C_\rho} \left| f(t) - P_n(t) \right|^{1/(1-\delta)} |dt| \right]^{1-\delta}$$

$$z \text{ interior to } C_\rho \; , \; 0 < \delta < 1.$$

This inequality is valuable in that it permits us to use a known degree of convergence by polynomials $P_n(z)$ to $f(z)$ on C_ρ in the sense of least weighted p-th powers to obtain results on the Tchebycheff degree of approximation of $p_n(z)$ to $f(z)$ on C. For future reference we state the theorem

THEOREM 5.1.3.   In Theorem 5.1.1 let

$$\left| \frac{\omega_n(z)}{\omega_n(t)} \right| \leqq M / \rho^n, \; z \text{ on } C, \; t \text{ on } C_\rho \; ,$$

$$\int_{C_\rho} \left| f(z) - P_n(t) \right|^{1/(1-\delta)} |dt| \leqq \epsilon_n, \; 0 < \delta < 1.$$

Then we have

$$\left| f(z) - p_n(z) \right| \leqq M_1 \, \epsilon_n^{1-\delta} / \rho^n \qquad z \text{ on } \overline{C}.$$

There are further theorems which might well be proved here but they follow without difficulty from the above results and are left as exercises for the reader.

The above theorems not only lead to the existence of polynomials converging to $f(z)$ on C with a certain degree of convergence, but also to the convergence properties of explicit polynomials found by interpolation to $f(z)$ in variously chosen sets of points z_j.

The results of this paragraph are due to Walsh and Sewell [1940].

§5.2. AN EXTENDED CLASSIFICATION. In our study of Problem α we considered functions of class $L(k, \alpha)$ on C for k a non-negative integer; we approximated the function $f(z)$ on C itself and hence we were not interested in functions which were not continuous on C. In Problem β, however, we study degree of approximation to $f(z)$ on C with relation to the behavior of $f(z)$ not on C but on or in a neighborhood of C_ρ; thus it is entirely appropriate under these circumstances to consider a much more general class of functions. We devote our attention first to the unit circle γ: $|z| = 1$.

A few preliminary remarks serve to motivate and accentuate the significance of the extended classification. If the function $f(z)$ belongs to the class $L(k, \alpha)$ on γ, $k \geq 0$, then the indefinite integral of $f(z)$ belongs to the class $L(k+1, \alpha)$ on γ, and for $k > 0$ the derivative $f'(z)$ of $f(z)$ belongs to the class $L(k-1, \alpha)$ on γ. These facts together with the necessary and sufficient condition of Theorem 1.2.1 form a natural basis for extending our original classification $L(k, \alpha)$ for $k \geq 0$ to negative (integral) values of k.

DEFINITION 5.2.1. If $f(z)$ is analytic in $|z| < 1$ and if we have

$$(5.2.1) \quad |f(re^{i\theta})| \leq L(1-r)^{k+\alpha}, \quad r < 1, \quad 0 < \alpha \leq 1,$$

where $k < 0$ is an integer, where $z = re^{i\theta}$, and where L is independent of r and θ, then we say that $f(z)$ belongs to the class $L(k, \alpha)$ on γ.

The following theorem indicates the connection between $k \geq 0$ and $k < 0$:

THEOREM 5.2.2. If the function $f(z)$ is of class $L(k, \alpha)$ on γ, $0 < \alpha \leq 1$, then the derivative $f'(z)$ is of class $L(k-1, \alpha)$ on γ, and the indefinite integral of $f(z)$ is of class $L(k+1, \alpha)$ on γ unless $\alpha = 1$, $k = -2$.

For $k > 0$ the conclusions of the theorem are well known (see Exs. 1.3.5 and 1.3.6); also for $k = 0$ the conclusion for the indefinite integral is immediate (see Ex. 1.3.6).

For the remaining values of k we consider first the derivative $f'(z)$. If $k = 0$ the conclusion follows from Theorem 1.2.1; if $k < 0$, so that $k + \alpha \leq 0$, the conclusion follows from Theorem 1.2.4.

For $k = -1$ the conclusion for the integral follows from Theorem 1.2.1; for $k < -1$, $k + \alpha \neq -1$, the conclusion follows from Theorem 1.2.5.

The conclusion on the integral is not valid for $\alpha = 1$ and $k = -2$, since in this case the proof of Theorem 1.2.5 fails. To take care of this exceptional case we introduce the following

DEFINITION 5.2.3. The function $f(z)$ belongs to the class $L'(k,1)$, $k \geq -1$, on γ provided $f^{(k+2)}(z)$ belongs to the class $L(-2,1)$.

We make the following observation:

THEOREM 5.2.4. If $f(z)$ is analytic and uniformly bounded in $|z| < 1$, then $f(z)$ belongs to the class $L'(-1,1)$.

As in the proof of Theorem 1.2.4 we have

$$f'(z_0) = \frac{1}{2\pi i}\int_{\delta}\frac{f(t)}{(t-z_0)^2}\,dt, \quad z_0 = re^{i\theta}, \quad r < 1,$$

where δ is the circle $|t - z_0| = \rho = (1-r)/2$; then

$$|f'(z_0)| \leq \frac{M}{2\pi}\,\frac{2\pi\rho}{\rho^2} = M\cdot\rho^{-1} = 2\,M(1-r)^{-1};$$

thus $f'(z)$ belongs to the class $L(-2,1)$, as was to be proved.

The following theorem indicates the close relation between functions of class $\mathrm{Log}(k,1)$ and $L'(k,1)$.

THEOREM 5.2.5. If $f(z)$ is of class $L'(k,1)$ on γ, $k > -1$, then $f'(z)$ is of class $L'(k-1,1)$ on γ; moreover $f^{(k+2+j)}(z)$ where j is a positive integer is of class $L(-2-j,1)$ on γ.

Also if $f(z)$ is of class $L'(k,1)$ on γ, then

$$(5.2.2) \quad \left|f^{(k+1)}(re^{i\theta})\right| \leq M_1|\log(1-r)|, \quad r < 1,$$

and $f(z)$ belongs to the class $\mathrm{Log}(k,1)$; here M_1 is independent of r and θ. Under these conditions the q-th integral of $f(z)$ is of class $L'(k+q,1)$, $q > 0$.

If $f(z)$ is of class $L'(k,1)$, $k \geq -1$, we have by definition $|f^{(k+2)}(z)| \leq M(1-r)^{-1}$; but $f^{(k+2)}(z)$ is the derivative of order $k + 1$ of $f'(z)$ and hence, also by the definition of class $L'(k,1)$, the function $f'(z)$ is of class $L'(k-1,1)$. Furthermore it is clear from the proof of Theorem 1.2.4 that $|f^{(k+2+j)}(z)| \leq M(1-r)^{-1-j}$, j a positive integer, and hence $f^{(k+2+j)}(z)$ belongs to the class $L\,(-2-j,1)$ by definition.

If $f(z)$ is of class $L'(k,1)$, an inequality on $f^{(k+1)}(z)$ follows directly from the inequality on $f^{(k+2)}(z)$:

$$\left|f^{(k+1)}(re^{i\theta}) - f^{(k+1)}(0)\right| = \left|\int_0^r f^{(k+2)}(re^{i\theta})dr\right|$$

$$\leq M\left|\int_0^r \frac{dr}{1-r}\right| = M|\log(1-r)|,$$

and since $f^{(k+1)}(0)$ is a constant we have the inequality of the theorem.

The function $f^{(k)}(z)$ can be defined on the boundary as the integral of its derivative and by integration along an arbitrary radius from r to 1 we obtain

$$(5.2.3)\quad \left|f^{(k)}(e^{i\theta}) - f^{(k)}(re^{i\theta})\right| \leq M(1-r)\,|\log(1-r)|,$$

$$r < 1.$$

Now let θ and θ' be given, $|\theta-\theta'| < \frac{1}{2}$ for simplicity. By (5.2.3) we have

$$\left|f^{(k)}(e^{i\theta}) - f^{(k)}(re^{i\theta})\right| \leq M(1-r)|\log(1-r)|$$

$$\left|f^{(k)}(e^{i\theta'}) - f^{(k)}(re^{i\theta'})\right| \leq M(1-r)|\log(1-r)|;$$

also by virtue of (5.2.2) we have

$$\left|f^{(k)}(re^{i\theta}) - f^{(k)}(re^{i\theta'})\right| \leq M|\log(1-r)|\cdot|\theta-\theta'|.$$

If we set $1-r = |\theta-\theta'|$ we see that $f(z)$ belongs to $\mathrm{Log}(k,1)$ on γ.

The remark about the q-th integral follows from the definition and the fact that the derivative of an indefinite integral is the function itself under the above conditions.

We extend our classification from the unit circle to an analytic Jordan curve by conformal mapping.

DEFINITION 5.2.6. Let Γ be an analytic Jordan curve in the z-plane. Let the interior of Γ be mapped conformally onto the interior of γ: $|w| = 1$, by the transformation $w = \Phi(z)$, $z = \Psi(w)$. The function $f(z)$ analytic interior to Γ is said to be of class $L(k, \alpha)$ or $L'(k, \alpha)$ on Γ if the function $f[\Psi(w)]$ (suitably defined on Γ if necessary) is of class $L(k, \alpha)$ or $L'(k, \alpha)$ respectively on γ, where $0 < \alpha \leq 1$ and where k is an integer, positive, negative, or zero.

Of course for $k \geq 0$ this coincides with the original definition of a function of class $L(k, \alpha)$. If Γ consists of several mutually exterior analytic Jordan curves we say $f(z)$ belongs to the class $L(k, \alpha)$ on Γ provided $f(z)$ belongs to the class $L(k, \alpha)$ on each component of Γ. For $k < 0$ the following theorem is an immediate consequence of Definition 5.4.1 and the analyticity of the Jordan curve.

THEOREM 5.2.7. Let Γ be an analytic Jordan curve, and let $\Gamma(\rho)$ be a sequence of analytic Jordan curves interior to Γ defined for all values of ρ in an interval $\rho_0 \leq \rho < \rho_1$ by an equation of the form

(5.2.4) $\Gamma(\rho) : |\Omega(z)| = \rho,$

where $\Omega(z)$ is analytic on Γ with $\Omega'(z) \neq 0$ on Γ, and with the property that $|\Omega(z)| = \rho_1$ on Γ. Then a necessary and sufficient condition that a function $f(z)$ belong to the class $L(k, \alpha)$, $k < 0$, on Γ is

(5.2.5) $|f(z)| \leq L(\rho_1 - \rho)^{k+\alpha}, \quad z \text{ on } \Gamma(\rho),$

where L is independent of z and ρ. The set Γ may consist of a finite number of mutually exterior analytic Jordan curves.

The property expressed by (5.2.4) and (5.2.5) is independent of the particular analytic function $\Omega(z)$ considered.

We leave to the reader the detailed statement and demonstration of the following extension:

THEOREM 5.2.8. Theorems 5.2.2, 5.2.4, and 5.2.5 are valid if the circle γ is replaced by several mutually exterior analytic Jordan curves Γ.

The extension of the classification $L(k, \alpha)$ to negative values of k and the related theorems included here are due to Walsh and Sewell [1941]; of course Theorems 1.2.1 to 1.2.5 due to Hardy and Littlewood [1932] form the basis for this classification.

§5.3. A POLYNOMIAL AND ITS INTEGRAL. In studying approximation in the sense of least weighted p-th powers for Problem β we need an estimation of the polynomial on C_ρ, rather than on C as in Problem α, implied by a bound on the integral over C of the p-th power of the polynomial.

We consider first a single curve for which we prove the following theorem:

THEOREM 5.3.1. Let C be a rectifiable Jordan curve or arc and let

$$(5.3.1) \qquad \int_C |P_n(z)|^p \, |dz| \leqq L^p, \quad p > 0,$$

where $P_n(z)$ is an arbitrary polynomial of degree n. Then we have

$$(5.3.2) \qquad |P_n(z)| \leqq KL\rho^n, \quad z \text{ on } \overline{C}_\rho,$$

where K is a constant depending on C, ρ, and p, but independent of $P_n(z)$, n, and z.

Let $\alpha_1, \alpha_2, \ldots, \alpha_\lambda$ be the zeros of $P_n(z)$ exterior to C, if there are no zeros we have $\lambda = 0$. Then the function

(5.3.3) $Q(z)$

$$= \frac{P_n(z)}{[\phi(z)]^n} \cdot \frac{[1-\overline{\phi}(\alpha_1)\phi(z)][1-\overline{\phi}(\alpha_2)\phi(z)]\ldots[1-\overline{\phi}(\alpha_\lambda)\phi(z)]}{[\phi(z)-\phi(\alpha_1)][\phi(z)-\phi(\alpha_2)]\ldots[\phi(z)-\phi(\alpha_\lambda)]}$$

is single valued and analytic (when suitably defined in the points α_ν and ∞) and is different from zero in the extended plane exterior to C; the second member of the right is replaced by 1 if $\lambda = 0$; here $\phi(z)$ is the usual mapping function. In (5.3.3) the second factor on the right removes the zeros of the $P_n(z)$ without affecting its modulus on C since $z = (1-\overline{\beta}w)/(w-\beta)$, $|\beta| > 1$, maps $|w| > 1$ on $|z| > 1$. Since C is a Jordan curve the function $Q(z)$ is continuous on C and we see that $|Q(z)| = |P_n(z)|$, z on C; thus

$$\int_C |Q(z)|^p \, |dz| \leq L^p .$$

The function $[Q(z)]^p/\phi(z)$ is analytic in the extended plane exterior to C, zero at infinity, and continuous on C, hence by the Cauchy integral formula we have

$$\frac{[Q(z)]^p}{\phi(z)} = \frac{1}{2\pi i} \int_C \frac{[Q(t)]^p \, dt}{\phi(t)(t-z)}, \quad z \text{ exterior to C.}$$

Since $|\phi(t)| = 1$, t on C, $|\phi(z)| = \rho$, z on C_ρ, $\rho > 1$, it follows that $|Q(z)| \leq KL$, z on C_ρ, where K depends on ρ but not on n nor on $Q(z)$. Also the function $|[\phi(z)-\phi(\alpha_\nu)]/[1-\overline{\phi}(\alpha_\nu)\phi(z)]|$ is unity for z on C, hence by its mapping property we know that it is less than unity for z on C_ρ. Thus we have $|P_n(z)| \leq |Q(z)|\,\rho^n$, z on C_ρ, and inequality (5.3.2) for z on C_ρ follows. An application of the principle of the maximum completes the proof.

Now let C consist of several mutually exterior analytic Jordan curves; the above proof does not generalize directly. We do consider, however, as above, the function $|P_n(z)/[\phi(z)]^n|^p$. This function is single-valued and continuous in the complement K of $\overline{C}$, even at infinity if suitably defined there; it is also subharmonic there, that is, it is not greater than the function, harmonic in K and continuous in K+C, which coincides with it on C. Hence for z in K we have

$$\left|\frac{P_n(z)}{[\phi(z)]^n}\right|^p \leqq \frac{1}{2\pi}\int_C \left|\frac{P_n(\zeta)}{[\phi(\zeta)]^n}\right|^p \frac{\partial G}{\partial \nu}\, ds$$

$$= \frac{1}{2\pi}\int_C |P_n(\zeta)|^p \frac{\partial G}{\partial \nu}\, ds,$$

where $G(z, \zeta)$ denotes the Green's function for K with pole at z, and ζ is the variable of integration. For z on C_ρ, the function $\partial G/\partial \nu$ is uniformly bounded on C, say not greater than L_1 in absolute value, so we may write from our hypothesis

$$\left|\frac{P_n(z)}{[\phi(z)]^n}\right| \leqq \frac{1}{2\pi} L_1\, L^p, \quad z \text{ on } C_\rho .$$

For z on C_ρ we have $|\phi(z)| = \rho$, thus the conclusion of the theorem follows by the principle of the maximum.

The requirement that C shall be composed of <u>analytic</u> Jordan curves can be lightened. It is clear that if $\partial G/\partial \nu$ is uniformly bounded and integrable for z on C_ρ and ζ on C the theorem is valid. Moreover let us suppose that C is composed of a finite number of rectifiable Jordan curves with the property that

$$\int_C \left|\frac{\partial G}{\partial \nu}\right|^{1/(1-\sigma)} ds, \quad 0 < \sigma < 1,$$

exists and is uniformly bounded, say not greater than L_2, for z on C_ρ. Then by the Hölder inequality (4.5.1) and by the subharmonic property of the function $|P_n(z)/[\phi(z)]^n|^{\sigma p}$, we may write

$$2\pi\left|\frac{P_n(z)}{[\phi(z)]^m}\right|^{\sigma p} \leqq \int_C \left|\frac{P_n(z)}{[\phi(z)]^m}\right|^{\sigma p} \frac{\partial G}{\partial \nu}\, ds$$

$$\leq \left[\int_C \left| \frac{P_n(z)}{[\phi(\zeta)]^m} \right|^p ds \right]^{\sigma} \left[\int_C \left| \frac{\partial G}{\partial \nu} \right|^{1/(1-\sigma)} ds \right]^{1-\sigma}$$

$$\leq L^{\sigma p} \, L_2^{1-\sigma}$$

The previous reasoning now suffices.

If C is a contour $\partial G/\partial \nu$ is continuous on C and is uniformly bounded for z on C_ρ; if C is a lemniscate or a line segment, some power greater than unity of $\partial G/\partial \nu$ is integrable on C and the integral is uniformly bounded for z on C_ρ.

Thus we have proved

THEOREM 5.3.2. Let C be a contour, a lemniscate, or a finite line segment, and let

$$\int_C |P_n(z)|^p \, |dz| \leq L^p, \quad p > 0,$$

where $P_n(z)$ is an arbitrary polynomial of degree n. Then we have

$$|P_n(z)| \leq KL \rho^n, \; z \text{ on } \overline{C}_\rho,$$

where K is a constant depending on C, ρ, and p, but independent of $P_n(z)$, n, and z.

Theorem 5.3.1 is due to Walsh [1935, pp. 92-93] and its extension to several curves, Theorem 5.3.2, to Walsh and Sewell [1940].

§5.4. EXERCISES. 5.4.1. Under the hypothesis of Theorem 5.1.1, the inequalities

$$|f(z) - P_n(z)| \leq \epsilon_n, \; z \text{ on } C_\rho,$$

$$\int_C |\omega_n(z)|^2 |dz| \leq \sigma_n, \; |\omega_n(t)| \geq \mu_n \text{ for } t \text{ on } C_\rho,$$

imply the inequality

$$\int_C |f(z) - p_n(z)|^2 |dz| \leq \frac{M_1 \sigma_n \epsilon_n^2}{\mu_n^2} .$$

[Walsh and Sewell, 1940]

5.4.2. Let C be an analytic Jordan curve, then

$$\left|\frac{F_n(z)}{F_n(t)}\right| \leq \frac{M}{\rho^n}, \ z \text{ on } C, \ t \text{ on } C_\rho ,$$

where $F_n(z)$ is the Faber polynomial of degree n belonging to C.

5.4.3. Let C be a Jordan curve such that

$$\frac{w \ \psi'(w)}{\psi(w) - \psi(w')} - \frac{w}{w-w'}$$

is bounded on $|w| = 1$ uniformly with respect to w' for $|w'| = 1$ and is expressible for $|w| > 1$ and $|w'| = 1$ by the Cauchy integral over $|w| = 1$. Then the conclusion of Ex. 5.4.2 is valid. [Walsh and Sewell, 1940]

5.4.4. Let C be a Jordan curve, let $\psi'(w)$ be different from zero on $|w| = 1$, and let $\psi''(w)$ exist and be continuous on $|w| = 1$. Then the conditions in Ex. 5.4.3 on C are fulfilled. [Walsh and Sewell, 1940]

5.4.5. Suppose that $|P_n(z)| \leq 1$ on the lemniscate $\Gamma : |z^2-1| = 1$, then $|P_n'(0)| \leq (en)^{1/2}$. (Compare Ex. 2.6.15)

5.4.6. Suppose that $|P_n(z)| \leq 1$ on the lemniscate $\Gamma : |z^m-1| = 1$, then $|P_n'(0)| \leq (en)^{1/m}$.

5.4.7. Let γ be the unit circle and suppose

$$\iint_{\overline{\gamma}} |P_n(z)|^2 \, d S \leq L^2.$$

Then $|P_n(z)| \leq KLn^{1/2}\rho^n$, $|z| \leq \rho > 1$, where K is a constant depending on ρ but not on $P_n(z)$, n, and z. The theorem is the best possible as far as the exponent of n is concerned. [Walsh]

5.4.8. Let γ be the unit circle and let

$$\iint_{\overline{\gamma}} |P_n(z)|^p \, dS \leq L^p , \quad p > 2.$$

Then $|P_n(z)| \leq KLn^{1/2}\rho^n$, $|z| \leq \rho > 1$, where K is a constant depending on ρ but not on $P_n(z)$, n, and z. [Walsh]

5.4.9. Let C be a Jordan curve composed of a finite number of analytic Jordan arcs meeting in corners of exterior openings less than or equal to $t\pi < 2\pi$; let

$$\iint_{\overline{C}} |P_n(z)|^p \, dS \leq L^p , \quad p > 0.$$

Then $|P_n(z)| \leq KLn^{2t/p}\rho^n$, $\rho > 1$, z on $\overline{C}_\rho$, where K is a constant independent of $P_n(z)$, n, z, and ρ.

§5.5. OPEN PROBLEMS. An obvious problem is the extension of Theorem 5.3.2 to sets bounded by arbitrary rectifiable Jordan curves. This would yield further results in the theory of approximation.

For surface integrals a theorem corresponding to Theorem 5.3.1 is totally lacking (compare Ex. 2.6.19 and the discussion concerning it in §2.7). For C an analytic Jordan curve and for p = 2 Carleman [1922] has established an asymptotic formula for the polynomials normal and orthogonal on $\overline{C}$; this formula is valid exterior to C and states that the polynomial of degree n is asymptotic to $n^{1/2}[\phi(z)]^n$; in this connection compare Ex. 5.4.7. The results of Exs. 5.4.8 and 5.4.9 are probably far from the best possible. In the conclusion of Ex. 5.4.8 the number p has completely disappeared; Ex. 5.4.7 and the asymptotic formula of Carleman lead one to suspect that p is the denominator in the exponent of n. In Ex. 5.4.9 the numerator in the exponent of n is 2t whereas in Ex. 5.4.7 it is t; this leads us to suspect that the inequality in the conclusion of Ex. 5.4.9 is indeed rough. The exact exponent of n is 1/2 for the unit circle which is either 1/p or t/p; it is conceivable that the exponent is $1/p$ for an arbitrary curve of type t; however this is purely a conjecture.

We might quite naturally study the functions $f(z)$ for which we have

$$|f(z)| \leqq L(\rho - r)^{k+\alpha}, \; k < 0, \; z \text{ on } C_r, \; 1 < r < \rho,$$

where C_ρ has multiple points. The reader will observe in what follows that certain results on degree of approximation to such functions are immediate. A thorough study of these functions would be of interest.

Chapter VI

TCHEBYCHEFF APPROXIMATION

§6.1. DIRECT THEOREMS. We are now ready to proceed with the study of degree of approximation so far as it concerns Problem β (see §1.1). Unlike the direct Tchebycheff theorems of Problem α whose proofs involved considerable machinery and manipulation, the direct Tchebycheff theorems of Problem β are relatively easy to prove thanks to the interpolation formula and inequalities of §5.1, and the inequalities of §2.5 on the equally distributed points.

Our fundamental result is

THEOREM 6.1.1. Let E be a closed limited set bounded by a contour C, and let C_ρ be a contour. Let the function $f(z)$ belong to the class $L(k, \alpha)$ on C_ρ, $\rho > 1$. Then $p_n(z)$ exists such that

$$(6.1.1)\quad |f(z) - p_n(z)| \leqq \frac{M}{n^{k+\alpha}\rho^n}, \quad z \text{ on } E.$$

For $k \geqq 0$ we have merely to take as the points of interpolation the equally distributed points on C, and in formula (5.1.2) to use as the polynomial $P_n(t)$ the polynomial of Theorem 3.2.1 which approximates to $f(z)$ on C_ρ; then $p_n(z)$ interpolates to $f(z)$ in n+1 equally distributed points on C. Under these conditions inequality (6.1.1) is an immediate consequence of Theorems 5.1.1, 2.5.7, and 3.2.1.

For $k < 0$ we use the definition of the class $L(k, \alpha)$ along with the classical interpolation formula. We have by formula (2.3.2)

$$f(z) - p_n(z) = \frac{1}{2\pi i} \int_{C_r} \frac{\omega_n(z)}{\omega_n(t)} \frac{f(t)}{(t-z)} dt \;, \quad z \text{ on } \overline{C}, \; 1 < r < \rho,$$

where the points of interpolation are equally distributed on C; then by virtue of Theorem 2.5.7 and Theorem 5.2.7 we have

$$|f(z) - p_n(z)| \leqq \frac{1}{2\pi} \frac{M_1}{r^n} \frac{M_2(\rho - r)^{k+\alpha}\, l}{(r-1)}, \quad z \text{ on } C,$$

where l is the length of C_ρ. If we set $r = \rho(1-1/n)$, it is clear that for n sufficiently large $r-1$ is uniformly bounded from zero; hence

$$|f(z) - p_n(z)| \leqq \frac{M_3\left(\frac{1}{n}\right)^{k+\alpha}}{\rho^n\left(1 - \frac{1}{n}\right)^n} \leqq \frac{M_4}{n^{k+\alpha}\, \rho^n}, \quad z \text{ on } C,$$

for n sufficiently large, say $n \geqq N$. Now to obtain (6.1.1) we have merely to choose an arbitrary but fixed sequence $p_n(z)$, $1 \leqq n < N$, and a constant M for which inequality (6.1.1) is valid for all n. The proof of the theorem is complete.

An extension of this theorem is immediate.

THEOREM 6.1.2. The conclusion of Theorem 6.1.1 is valid for $k \geqq 0$ if C is a lemniscate provided C_ρ is a contour.

The proof of Theorem 6.1.1 for $k \geqq 0$ is valid for Theorem 6.1.2 since C_ρ consists of mutually exterior analytic Jordan curves, thus inequality (2.4.5) and Theorem 3.2.1 apply.

The significance of the above theorems is two-fold: (i) we have established existence theorems for sequences converging with at least a certain degree of convergence, (ii) we have determined inequalities for the convergence of specific sequences of polynomials. In the statements of the theorems we have mentioned only item (i); many results concerning item (ii) are included in the exercises at the end of this chapter.

For $k \geq 0$ the results of this section are due to Walsh and Sewell [1940]; the same authors later [1941] extended Theorem 6.1.1 to include negative values of k.

§6.2. OPERATIONS WITH APPROXIMATING SEQUENCES. Before proceeding with the indirect theorems corresponding to the direct theorems of the preceding section we consider the effect upon degree of approximation of the operations of integration and differentiation (compare §3.5). These results are not only interesting for their own sakes but also for obtaining new direct theorems; we shall by this method extend Theorem 6.1.1 to include functions of class $L'(k, \alpha)$ and present a new proof of this theorem for $k \geq 0$.

We consider first the unit circle and the Taylor development.

THEOREM 6.2.1. Let $f(z)$ belong to the class $L(k, \alpha)$, $k \leq -1$, $0 < \alpha \leq 1$, on γ: $|z| = 1$; let $p_n(z)$ denote the sum of the first n+1 terms of the Taylor development of $f(z)$. Then we have

$$(6.2.1) \quad \left| \int_0^z [f(z) - p_n(z)]dz \right| \leq \frac{M}{n^{k+\alpha+1}\rho^n},$$

$|z| \leq 1/\rho < 1$.

By virtue of the interpolating property of the Taylor development we have

$$f(z)-p_n(z) = \frac{1}{2\pi i} \int_{|t|=r} \frac{z^{n+1} f(t)\, dt}{t^{n+1}(t-z)}, \quad |z| \leq 1/\rho < r < 1,$$

and hence

$$\int_0^z [f(z) - p_n(z)]dz = \frac{1}{2\pi i} \int_{|t|=r} \frac{f(t)}{t^{n+1}} dt \int_0^z \frac{z^{n+1}}{t-z} dz,$$

where for simplicity the path of integration is chosen along a radius. But for $|t| > |z|$ we have

$$\int_0^z \frac{z^{n+1}}{t-z} dz = \frac{1}{t} \left[\frac{z^{n+2}}{n+2} + \frac{z^{n+3}}{(n+3)t} + \cdots \right] ;$$

for $|t| = r$ and $|z| \leq 1/\rho$ the modulus of this function is dominated by

$$\frac{1}{nr\rho^n} \left[\frac{1}{1 - \frac{1}{r\rho}} \right].$$

Thus by the method employed in the proof of Theorem 6.1.1 for $k < 0$ we obtain inequality (6.2.1).

Theorem 6.2.1 is stated merely for the first integral of a function of class $L(k, \alpha)$, but obviously extends to the iterated indefinite integrals of every order. This theorem yields a new proof of Theorem 6.1.1 for C_ρ the unit circle (lemniscate) and $k \geq 0$, by virtue of Theorem 5.2.2. Also for functions of class $L'(k,1)$, $k \geq -1$, we have

THEOREM 6.2.2. If $f(z)$ belongs to the class $L'(k,1)$, $k \geq -1$, on γ: $|z| = 1$, then $p_n(z)$ exists such that

$$(6.2.2) \quad |f(z)-p_n(z)| \leq \frac{M}{n^{k+1} \rho^n}, \quad |z| \leq 1/\rho < 1.$$

By definition $f^{(k+2)}(z)$ belongs to the class $L(-2,1)$, hence by Theorem 6.1.1 the hypothesis of Theorem 6.2.1 is fulfilled; repeated application of the conclusion of Theorem 6.2.1 then yields inequality (6.2.2). Here $p_n(z)$ is the sum of the first $n + 1$ terms of the Taylor development of the function $f(z)$.

We now consider a set bounded by an analytic Jordan curve; the following theorem is stated merely for the first integral but also extends at once to higher integrals.

THEOREM 6.2.3. Let C be an analytic Jordan curve and let $f(z)$ belong to the class $L(k, \alpha)$, $k \leq -1$, $0 < \alpha \leq 1$, on C_ρ. Let $p_n(z)$ denote the polynomial of degree n defined by interpolation to $f(z)$ in points equally distributed on a suitable level curve $C_{1-\delta}$, $\delta > 0$, interior to C belonging to the analytic family of curves C_σ. Then we have

$$(6.2.3) \quad \left| \int_a^z [f(z)-p_n(z)]dz \right| \leq \frac{M}{n^{k+\alpha+1}\rho^n}, \quad z \text{ on } \overline{C},$$

where a is an arbitrary point interior to C and where the path of integration contains no point exterior to C.

We use the interpolation formula (2.3.2)

$$f(z)-p_n(z) = \frac{1}{2\pi i}\int_{C_r} \frac{\omega_n(z)f(t)dt}{\omega_n(t)(t-z)}, \quad 1 < r < \rho;$$

for z on or within C we have

$$\int_a^z [f(z)-p_n(z)]dz = \frac{1}{2\pi i}\int_{C_r} \frac{f(t)dt}{\omega_n(t)} \int_a^z \frac{\omega_n(z)\,dz}{t-z} .$$

Now let $\delta > 0$ be chosen so small that the locus $C_{1-\delta}$: $|\phi(z)| = 1-\delta$ is an analytic Jordan curve interior to C, where $\phi(z)$ is the usual mapping function; we take the points z_j of interpolation, defining $\omega_n(z)$, equally distributed on $C_{1-\delta}$. Let $\zeta = w/(1-\delta) = \phi(z)/1-\delta)$ and $e^g = (1-\delta)/\phi'(\infty) = (1-\delta)\Delta$; thus $\zeta = \phi(z)/(1-\delta)$ maps the exterior of $C_{1-\delta}$ onto $|\zeta| > 1$. We have by (2.5.30)

$$e^{-M} \leq \left| \frac{\omega_n(z)}{e^{(n+1)g}(\zeta^{n+1}-1)} \right| \leq e^M ,$$

for z on or exterior to $C_{1-\delta}$. Consequently for z on $C_{1-\delta}$ we have

$$| \omega_n(z)| \leq M_1 \Delta^{n+1} \cdot (1-\delta)^{n+1}.$$

The function $\omega_n(z)/\zeta^{n+1}$ is analytic in the closed exterior of $C_{1-\delta}$ even at infinity, and ζ has the modulus unity for z on $C_{1-\delta}$, so we may write for z on or exterior to $C_{1-\delta}$

$$(6.2.4) \quad | \omega_n(z)| \leq M_2 \Delta^{n+1} \cdot (1-\delta)^{n+1} | \zeta |^{n+1} .$$

We integrate from an arbitrary z_0 on $C_{1-\delta}$ to z on C, choosing as path the image ("Radiusbild") in the z-plane of a radius of the unit circle in the plane of $\zeta = \phi(z)/(1-\delta)$; for all t on C_r where r is sufficiently near ρ , we have

$$\left| \int_{z_0}^{z} \frac{\omega_n(z)dz}{t-z} \right| \leq M_3 \int_{z_0}^{z} \Delta^{n+1} |\phi(z)|^{n+1} |dz|$$

$$\leq M_4 \int_{z_0}^{z} \Delta^{n+1} |\phi(z)|^{n+1} |\phi'(z)dz|$$

$$= M_4 \Delta^{n+1} [|\phi(z)|^{n+2} - |\phi(z_0)|^{n+2}]/(n+2)$$

$$= M_5 \Delta^{n+2}/(n+2).$$

Here M_5 is independent of n, z_0, and z. If a is a fixed point interior to $C_{1-\delta}$, further use of (6.2.4) yields

$$\left| \int_{a}^{z} \frac{\omega_n(z)dz}{t-z} \right| \leq \left| \int_{a}^{z_0} \frac{\omega_n(z)dz}{t-z} \right| + \left| \int_{z_0}^{z} \frac{\omega_n(z)dz}{t-z} \right| \leq \frac{M_6 \Delta^{n+2}}{n+2} .$$

Also for t on or exterior to C we have $| \omega_n(t)| \geq M_7 \Delta^{n+1} |\phi(t)|^{n+1} > 0$; hence for z on C we have

$$\left| \int_{\alpha}^{z} [f(z)-p_n(z)]dz \right| \leq M_8 \cdot \max [|f(t)|, t \text{ on } C_r]/r^n n.$$

Inequality (6.2.3) is now obtained by the method of proof of Theorem 6.1.1 for $k < 0$, Theorem 6.2.3 is now established.

The reasoning cannot be carried through if the points of interpolation are chosen equally distributed on C itself, as we now indicate. Let C be the unit circle γ: $|z| = 1$, whence $\omega_n(z) \equiv z^{n+1} - 1$; then for z on C we have

$$\int_0^z (z^{n+1} - 1)\, dz = \frac{z^{n+2}}{n+2} - z,$$

so no additional factor n appears in the denominator due to the integration.

We are now in a position to formulate an analogue of Theorem 6.2.2 which follows from the above result just as Theorem 6.2.2 follows from Theorem 6.1.1.

THEOREM 6.2.4. Let C be an analytic Jordan curve and let f(z) belong to the class $L'(k,1)$, $k \geq -1$, on C_ρ. Then $p_n(z)$ exists such that

$$|f(z) - p_n(z)| \leq \frac{M}{n^{k+1}\rho^n}, \quad z \text{ on } \overline{C}.$$

This is an interesting complement to Theorem 6.1.1; it is to be observed that we establish the same degree of approximation for a function of class $L'(k,1)$ here as we did in Theorem 6.1.1 for a function of class $L(k,1)$.

We now state a more general theorem for continuous functions.

THEOREM 6.2.5. Let C be an analytic Jordan curve. Let f(z) be analytic interior to C_ρ and continuous on $\overline{C}_\rho$, and let $P_n(z)$ exist such that we have

$$|f(z) - P_n(z)| \leq \epsilon_n, \; z \text{ on } C_\rho .$$

Let $p_n(z)$ denote the polynomial which interpolates to $f(z)$ in $n+1$ equally distributed points on $C_{1-\delta}$, $\delta > 0$. Then we have

$$\left| \int_a^z [f(z) - p_n(z)]dz \right| \leq \frac{M\epsilon_n}{n\rho^n},$$

where a is an arbitrary point interior to C, and where the integral is taken along an arbitrary path containing no point exterior to C.

The proof here is similar to that of the preceding theorem. Instead of formula (2.3.2) we use formula (5.1.2). The details are left to the reader. It will be noticed that Theorem 6.2.5 with its extension to higher integrals affords a new proof of Theorem 6.1.1 for $k \geq 0$ by virtue of Theorem 3.2.1 for $k = 0$.

We prove corresponding results for differentiation.

THEOREM 6.2.6. Let C be an analytic Jordan curve; let $f(z)$ belong to the class $L(k, \alpha)$, $k \leq -1$, $0 < \alpha \leq 1$, on C_ρ. Let $p_n(z)$ denote the polynomial which interpolates to $f(z)$ in $n+1$ equally distributed points on C. Then we have

$$|f'(z) - p_n'(z)| \leq \frac{M}{n^{k+\alpha-1}\rho^n}, \; z \text{ on } C.$$

In formula (2.3.2) let us differentiate with respect to z

$$f'(z)-p_n'(z) = \frac{1}{2\pi i}\int_{C_r} \frac{f(t)}{\omega_n(t)}\left[\frac{(t-z)\,\omega_n'(t)+\omega_n(z)}{(t-z)^2}\right]dt, \quad 1<r<\rho .$$

But $|\omega_n(z)| \leq M\Delta^{n+1}$, z on C, and hence by Theorem

2.1.4 it follows that $|\omega_n'(z)| \leqq M\,\Delta^{n+1}(n+1)$, z on C. Thus we have

$$(6.2.5)\quad |f'(z)-p_n'(z)| \leqq \frac{M_1 n \cdot \max\,[|f(t)|,\ t \text{ on } C_r]}{r^{n+1}}\ ;$$

theorem (6.2.6) now follows by the method of proof of Theorem 6.1.1 for $k < 0$.

For functions continuous on C_ρ we have

THEOREM 6.2.7. Let C be an analytic Jordan curve, and let $f(z)$ be analytic interior to the analytic Jordan curve C_ρ, continuous in the corresponding closed region. Let $P_n(z)$ exist such that

$$|f(z) - P_n(z)| \leqq \epsilon_n,\ z \text{ on } C_\rho .$$

Let $p_n(z)$ denote the polynomial which interpolates to $f(z)$ in $n+1$ equally distributed points on C. Then we have

$$|f'(z) - p_n'(z)| \leqq M\,\epsilon_n\, n/\rho^n,\ z \text{ on } C.$$

The proof here is similar to that of the preceding theorem except that we use formula (5.1.2). The details are left to the reader.

Theorem 6.2.7 by virtue of Theorem 3.2.1 includes the conclusion of Theorem 6.2.6 for $k \geqq 0$, $0 < \alpha \leqq 1$; in case $k + \alpha = 0$ we set $\epsilon_n = M_0$.

Theorems 6.2.6 and 6.2.7 extend to a single Jordan curve which is a contour.

Corresponding to Theorem 3.4.2 of Problem α we have for Problem β the theorem

THEOREM 6.2.8. Let C be a Jordan arc or curve of Type t and let $f(z)$ be defined on $\overline{C}$. Let $p_n(z)$ exist such that

$$(6.2.6)\quad |f(z)-p_n(z)| \leqq \frac{M}{n^{\beta t}\,\rho^n},\ z \text{ on } \overline{C},\ \rho > 1.$$

Then f(z) is analytic on $\overline{C}$, and

(6.2.7) $$|f'(z) - p_n'(z)| \leq \frac{M_1}{n^{(\beta-1)t}\rho^n}, \quad z \text{ on } C.$$

As usual we write

(6.2.8) $$f(z) = p_1(z) + [p_2(z)-p_1(z)] + \dots + [p_{n+1}(z) - p_n(z)] + \dots ;$$

we have by virtue of (6.2.6) for z on $\overline{C}$

$$|p_{n+1}(z) - p_n(z)| \leq \frac{2M}{n^{\beta t}\rho^n}, \quad \beta \geq 0,$$

$$|p_{n+1}(z) - p_n(z)| \leq \frac{2M}{(n+1)^{\beta t}\rho^n}, \quad \beta < 0,$$

Thus by Theorem 2.1.4 we have then for $\beta \geq 1$

$$|p_{n+1}'(z) - p_n'(z)| \leq \frac{M_1}{n^{(\beta-1)t}\rho^n}, \quad z \text{ on } C;$$

hence from (6.2.8) we write

$$|f'(z) - p_n'(z)| \leq M_1 \sum_{\nu=n}^{\infty} \frac{1}{\nu^{(\beta-1)t}\rho^\nu},$$

whence inequality (6.2.7) is immediate for $\beta \geq 1$. If $\beta < 1$ we choose ρ_1, $1 < \rho_1 < \rho$, and the inequality

$$\left[\frac{n+1}{n+2}\right]^{(\beta-1)t} < \rho/\rho_1$$

is valid for sufficiently large n. Then for sufficiently large n we have

$$\frac{n+1}{n+\nu} = \frac{n+1}{n+2} \cdot \frac{n+2}{n+3} \cdots \frac{n+\nu-1}{n+\nu} \leq \left(\frac{n+1}{n+2}\right)^{\nu-1}$$

Consequently we obtain

$$|f'(z)-p_n'(z)| \leq \frac{M_1}{(n+1)^{(\beta-1)t}\rho^n}\left[1+\left(\frac{n+1}{n+2}\right)^{(\beta-1)t}\frac{1}{\rho} + \left(\frac{n+1}{n+3}\right)^{(\beta-1)t}\frac{1}{\rho^2} + \dots\right]$$

$$\leq \frac{M_1}{(n+1)^{(\beta-1)t}\rho^n} \left[1 + \frac{1}{R_1} + \frac{1}{R_1^2} + \cdots\right]$$

$$\leq \frac{M_2}{n^{(\beta-1)t}\rho^n}, \quad z \text{ on } E.$$

This holds for all n by a suitable choice of M_2 and the proof of the theorem is complete.

Theorem 6.2.8 includes Theorem 6.2.6 by virtue of Theorem 6.1.1. It is clear that the conclusion of Theorem 6.2.8 is valid with $t = 2$ for an arbitrary set whose complement is simply connected.

Theorems 6.2.1 to 6.2.7 inclusive are due to Walsh and Sewell [1941].

§6.3. INDIRECT THEOREMS. We now consider theorems corresponding to the direct theorems of §§6.1 and 6.2 but in the converse direction. Our method consists first in obtaining from the original inequality on C a bound on C on the polynomials involved and applying Theorem 2.1.3 to establish an inequality on C_r, $1 < r$. For $k \geq 0$ we could then establish a degree of approximation to $f(z)$ on C_ρ and apply Theorem 3.2.1; however, by using Theorem 6.2.8 after differentiation and then applying Theorem 5.2.8 or Definitions 5.2.3 and 5.2.6, we obtain a proof which is valid for k negative as well as non-negative.

THEOREM 6.3.1. Let C be an analytic Jordan curve and let $f(z)$ be defined on $\overline{C}$. Let $p_n(z)$ exist such that

$$(6.3.1) \quad |f(z) - p_n(z)| \leq \frac{M}{n^{k+\alpha+1}\rho^n}, \quad z \text{ on } \overline{C}, \ \rho > 1.$$

Then $f(z)$ when suitably defined exterior to E belongs to the class $L(k, \alpha)$ on C_ρ if $k+\alpha+1$ is not a positive integer and to the class $L'(k, \alpha)$ if $k+\alpha+1$ is a positive integer.

The proof follows that of Theorem 6.2.8. Suppose first that $k+\alpha < 0$; we write $k+\alpha+1 = q$, then as in the proof of Theorem 6.2.8 we have by (6.3.1)

$$|p_{n+1}(z) - p_n(z)| \leqq \frac{2M}{(n+1)^q \rho^n}, \quad z \text{ on } C.$$

Consequently

$$|p_{n+1}(z)-p_n(z)| \leqq \frac{M_1 r^{n+1}}{(n+1)^q \rho^n} \quad z \text{ on } C_r, \; 1 < r < \rho,$$

by Theorem 2.1.3; the analyticity of $f(z)$ in C_ρ is an immediate consequence of this second inequality. Hence

$$|f(z)| \leqq M_2 \sum_{\nu=2}^{\infty} \frac{r^\nu}{\nu^q \rho^\nu} \leqq M_2 \int_0^\infty \left(\frac{r}{\rho}\right)^x \cdot x^{-q}\, dx$$

$$= M_2 \int_0^\infty e^{x \log \frac{r}{\rho}} x^{-q}\, dx$$

$$= M_2 \Gamma(1-q)\left(-\log \frac{r}{\rho}\right)^{-q+1} \qquad \leqq M_3(1-r/\rho)^{q-1}$$

$$\leqq M_4(\rho-r)^{q-1}, \quad z \text{ on } C_r.$$

Thus we have

$$|f(z)| \leqq M_4(\rho-r)^{k+\alpha}, \quad z \text{ on } C_r,$$

and hence by Theorem 5.2.7 the function $f(z)$ belongs to the class $L(k, \alpha)$ on C_ρ.

Now suppose $k+\alpha \geqq 0$; we define p as the positive integer such that $k+\alpha < p \leqq k+\alpha+1$. Then by Theorem 6.2.8 it follows that

$$\left|f^{(p)}(z) - p_n^{(p)}(z)\right| \leqq \frac{M_1}{n^{k+\alpha+1-p} \rho^n},$$

and if we set $q = k + \alpha + 1-p$ we see as above that $f^{(p)}(z)$ belongs to the class $L(k-p, \alpha)$ on C_ρ. If $k+\alpha$ is not zero or a positive integer we have merely to integrate p times applying Theorem 5.2.8 to prove that $f(z)$ belongs to the class $L(k, \alpha\ )$ on C_ρ. If $k+\alpha$ is zero or a positive integer we see from above that $k-p = -2$, $\alpha = 1$ and hence $f^{(p)}(z)$ belongs to the class $L(-2,1)$ on C_ρ; consequently by Definitions 5.2.3 and 5.2.6 we see that $f(z)$ belongs to the class $L'(p-2,1)$ or to the class $L'k, \alpha)$. The proof of the theorem is complete.

By virtue of the fact that Theorem 6.2.8 is valid for C a Jordan curve which is a contour the above theorem holds also for this case; we recall (§2.5) that $t = 1$ for a contour. As a matter of fact by applying Theorem 2.1.4 to $|p_{n+1}(z)-p_n(z)|$ for z on C_r and showing that for $1 < r_o < r \leqq \rho$ the constant $K(C_r)$ in Theorem 2.1.4 is independent of r, Theorem 6.3.1 can be extended to an arbitrary set E whose complement is simply connected.

We also have for $k \geqq 0$ the following theorem

THEOREM 6.3.2. Let E, with boundary C, be a closed limited set whose complement is connected and regular and let $f(z)$ be defined on E. Let $p_n(z)$ exist such that

$$(6.3.2)\quad |f(z)-p_n(z)| \leqq \frac{M}{n^{k+\alpha+1}\rho^n},\quad z \text{ on } E,$$

$k \geqq 0$, $0 < \alpha \leqq 1$.

Then if $0 < \alpha < 1$ the function $f(z)$ belongs to the class $L(k, \alpha)$ on C_ρ, and if $\alpha = 1$ to the class $Log(k,1)$ on C_ρ.

We define $f(z)$ as usual by the series and from (6.3.2) we have

$$|p_{n+1}(z)-p_n(z)| \leqq \frac{M_1}{n^{k+\alpha+1}\rho^n},\quad z \text{ on } C,$$

$$|p_{n+1}(z)-p_n(z)| \leqq \frac{M_2}{n^{k+\alpha+1}},\quad z \text{ on } C_\rho .$$

Hence

$$|f(z)-p_n(z)| \leqq M_2 \sum_{\nu=n}^{\infty} \frac{1}{\nu^{k+\alpha+1}} \leqq \frac{M_3}{n^{k+\alpha}},\quad z \text{ on } C_\rho .$$

If C_ρ consists of mutually exterior analytic Jordan curves we have merely to apply Theorem 3.4.1 directly. However, as mentioned in connection with the proof of Theorem 3.4.1, inequality (3.4.2) is a property in the small and each properly chosen branch of C_ρ is analytic even across an intersection, hence inequality (3.4.2) can be established everywhere on C_ρ by considering overlapping intervals.

To show that $\alpha = 1$ is exceptional both in Theorems 6.3.1 and 6.3.2 we use an example similar to that employed in connection with Theorem 3.4.1. Let C be the circle $|z| = 1/\rho < 1$ and C_ρ the unit circle and consider

$$f(z) = \sum_{\nu=2}^{\infty} \frac{z^\nu}{\nu^q},$$

where q is a positive integer. On $|z| = 1/\rho$ we have for $q > 0$

$$\left| f(z) - \sum_{\nu=2}^{n} \frac{z^\nu}{\nu^q} \right| \leqq \sum_{\nu=n+1}^{\infty} \frac{1}{\nu^q \rho^\nu}$$

$$\leqq \frac{M}{n^q \rho^n}, \quad |z| \leqq 1/\rho .$$

On the other hand if $q = 1$ we have

$$f(z) = \sum_{\nu=2}^{\infty} \frac{z^\nu}{\nu} = \log (1-z) - z;$$

hence $f(z)$ is not bounded in $|z| < 1$ and consequently does not belong to the class $L(-1,1)$ on $|z| = 1$; if $q = 2$ we likewise see that the derivative $f'(z)$ becomes infinite as z approaches 1 and hence $f(z)$ cannot belong to the class $L(0,1)$ on $|z| = 1$; it belongs to the class $\mathrm{Log}(0,1)$, of course. This argument extends easily to show that $\alpha = 1$ is exceptional for arbitrary positive integral values of k.

Theorem 6.3.1 is proved by Walsh and Sewell [1941] by a different method. Theorem 6.3.2 is due to Walsh and Sewell [1937a] for C_ρ consisting of a finite number of mutually exterior Jordan curves.

§6.4. COUNTER EXAMPLES. A comparison of the direct theorems of §§6.1 and 6.2 and the indirect theorems of §6.3 shows that they are not exact converses; in fact there is a discrepancy of unity in the exponents of n. This discrepancy is inherent in the nature of the problem, as we shall now show by examples.

We consider first the direct theorems, using the example (4.1.2) of §4.1. Here we let C be the circle $|z| = 1/\rho$, setting $\rho = 2$ for definiteness, and thus C_ρ is the unit circle $|z| = 1$. Let $k = 0$, $0 < \alpha < 1$; then we have already shown that $f(z)$ belongs to the class $L(0, \alpha)$ on C_ρ. On the other hand let $L_n(z)$ be the polynomial of degree n of best approximation to $f(z)$ on C in the sense of Tchebycheff; we set

$$\epsilon_n = \max\,[|f(z) - L_n(z)|,\ z \text{ on } C].$$

The polynomial $S_n(z)$ (in the notation of §4.1) is the polynomial of degree n of best approximation to $f(z)$ on C in the sense of least squares, so we have

$$2\pi\epsilon_n^2 \geqq \int_C |f(z) - L_n(z)|^2 |dz| \geqq \int_C |f(z) - S_n(z)|^2 |dz|.$$

For the particular value $n = m_{j+1} - 1$ this yields for j sufficiently large

$$\epsilon_n^2 \geqq \sum_{\nu=j+1}^{\infty} \frac{1}{2^{2m_\nu} \cdot 2^{2\nu}} > \frac{1}{2^{2m_{j+1}} \cdot 2^{2j+2}}$$

$$(6.4.1) \qquad = \frac{1}{2^6 2^{2m_{j+1}-2}\ 2^{2j-2}},$$

$$\epsilon_n > \frac{1}{8 \cdot 2^n \cdot n^\alpha}.$$

That is to say, the inequality (6.4.1) obtains for an infinite sequence of indices n, so for the function (4.1.2) it is not possible to conclude an inequality of the form

$$|f(z) - L_n(z)| \leq \frac{M_1}{n^{\alpha'} \rho^n}, \quad z \text{ on } C,$$

where $\alpha' > \alpha$, or even an inequality of the form

$$|f(z) - L_n(z)| \leq \frac{\delta(n)}{n^{\alpha} \rho^n}, \quad z \text{ on } C,$$

where $\delta(n)$ approaches zero as n becomes infinite. Thus the direct theorems for the case $k = 0$, $0 < \alpha < 1$, cannot be improved in the sense mentioned.

For class $L(0,1)$ the argument is essentially the same as that used in the corresponding case in §4.1; only obvious changes are necessary. Thus we have shown that the direct theorems for the case $k = 0$, $0 < \alpha \leq 1$, cannot be improved in the sense of increasing the exponent of n. Since the integral of the Taylor development of a function is the Taylor development of the integral of the function the extension of this remark to positive values of k follows from Theorems 6.1.1, 6.2.1, and 5.2.2, and Definition 5.2.3; the extension to negative values of k follows likewise from Theorem 6.2.8 by differentiation, since the derivative of the Taylor development of a function is the Taylor development of the derivative of the function. The details of these extensions are left to the reader.

We turn now to the indirect theorems. The example already exhibited in connection with Theorems 6.3.1 and 6.3.2 takes care of the case $k \geq -1$, $\alpha = 1$, in that it shows that these theorems cannot be improved in the sense that the conclusions become false if we replace α in inequalities (6.3.1) and (6.3.2) by any particular $\alpha' < \alpha$. For the remaining values of k and α we consider the function (see §4.2)

$$f(z) \equiv (1-z)^{\beta} \equiv \sum_{\nu=0}^{\infty} a_{\nu} z^{\nu},$$

where β is not a positive integer; we know from the results of §4.2 that

$$|a_n| \leqq M/n^{\beta+1} .$$

Hence

$$\left| f(z) - \sum_{\nu=0}^{n} a_\nu z^\nu \right| \leqq \sum_{\nu=n+1}^{\infty} \frac{M}{\nu^{\beta+1} \rho^\nu} , \quad |z| \leqq 1/\rho ,$$

thus it follows that

$$\left| f(z) - \sum_{\nu=0}^{n} a_\nu z^\nu \right| \leqq \frac{M_1}{n^{\beta+1} \rho^n} , \quad |z| \leqq 1;$$

for β non-negative this inequality is immediate and for β negative we use the argument employed in the proof of Theorem 6.2.8. If $\beta = k + \alpha$, $0 < \alpha < 1$, $k = 0$, we have shown that $f(z)$ belongs to the class $L(0, \alpha)$ on $|z| = 1$ and to no higher class; here we take, as usual, C_ρ as the unit circle and C as the circle $|z| = 1/\rho < 1$. For $k > 0$ the function $f(z)$ belongs to the class $L(k, \alpha)$ and to no higher class by the same argument as before. For $k = -1$, $0 < \alpha < 1$, or for $k < -1$, $0 < \alpha \leqq 1$, we see by definition that $f(z)$ belongs to the class $L(k, \alpha)$ and to no higher class. Thus the indirect theorems cannot be improved in the sense mentioned.

The examples employed in the present section are used by Walsh and Sewell [1940,1941] in the same connection.

§6.5. FUNCTIONS WITH ISOLATED SINGULARITIES. Hitherto we have considered functions of class $L(k, \alpha)$ or $L'(k, \alpha)$ on C_ρ ; these functions satisfy given continuity or asymptotic conditions on or in the neighborhood of C_ρ uniformly with respect to every point of C_ρ . The class of a function is determined by its behavior at (or in the neighborhood of) its worst point, so to speak. For example, a function which is analytic at every point of $\overline{C}_\rho$ except at the point z_0, in the neighborhood of which it satisfies a Lipschitz condition of order α and no higher order, belongs to the class $L(0, \alpha)$ on C_ρ and to no higher class; on the other hand a function which is analytic in C_ρ, has C_ρ as a cut, and satisfies at each point of C_ρ a Lipschitz condition of order α and no higher order belongs likewise to the same class $L(0, \alpha)$ on C_ρ. Our direct theorems (§§6.1 and 6.2) put these functions in the same category as far as degree of approximation is concerned; however, we saw in §6.4 that for the function* $f(z) = (1-z)^{k+\alpha} = \sum_{\nu=0}^{\infty} a_\nu z^\nu$ we have

$$(6.5.1) \qquad \left| f(z) - \sum_{\nu=0}^{n} a_\nu z^\nu \right| \leqq \frac{M}{n^{k+\alpha+1}\rho^n}, \quad |z| \leqq 1/\rho < 1,$$

and yet for $0 < \alpha < 1$ the function $f(z)$ belongs to the class $L(k, \alpha)$ on C_ρ : $|z| = 1$, and to no higher class. Thus we obtain for this function a higher degree of approximation than is asserted in the direct theorems; it is quite natural then to pursue further the study of degree of approximation for such functions, that is functions with isolated singularities. This section is devoted to just such functions.

We state a generalization of the above conclusion:

THEOREM 6.5.1. Suppose

$$f(z) = F_1(z) + F_2(z) + \cdots + F_\sigma(z)$$
$$+ a_1(z-z_1)^{h_1} + \cdots + a_\lambda (z-z_\lambda)^{h_\lambda},$$
$$|z_j| = 1, \quad j = 1, 2, \ldots, \lambda,$$

where F_i, $i = 1, \ldots, \sigma$, belongs to the class $L(k_i, \alpha_i)$ or $L'(k_i, \alpha_i)$ on $|z| = 1$. Let $H_i = k_i + \alpha_i$ and $h = \min(h_j - 1, H_i)$.
Then $p_n(z)$ exists such that

$$|f(z) - p_n(z)| \leqq \frac{M}{n^h \rho^n}, \quad |z| \leqq 1/\rho < 1.$$

* Compare Dienes [1913]; Mandelbrojt [1927,1932].

The proof simply consists in applying Theorem 6.1.1 or Theorem 6.2.2, as the case may be, and the conclusion (6.5.1) to the appropriate component parts of the function f(z); of course the points z_j need not be distinct. If here $f(z) = \sum a_\nu z^\nu$ we have $|a_\nu| \leq M_1/\nu^h$.

In a similar way we obtain for the special function $f(z) = \log(1-z)$, which belongs to the class $L'(-1,1)$ a stronger result than that of Theorem 6.2.2.

THEOREM 6.5.2. Suppose $f(z) = \log(1-z) = \sum_{\nu=1}^{\infty} z^\nu/\nu$, $|z| < 1$. Then we have

$$\left| f(z) - \sum_{\nu=1}^{\infty} z^\nu/\nu \right| \leq \frac{M}{n\rho^n}, \quad |z| \leq 1/\rho < 1.$$

These theorems can be extended to somewhat more general functions by means of certain inequalities on multiplication of series [see, e.g., Hardy and Littlewood, 1935]. Let $f(z) = \sum_{\nu=0}^{\infty} a_\nu z^\nu$, $g(z) = \sum_{\nu=0}^{\infty} b_\nu z^\nu$, $f(z)g(z) = \sum_{\nu=0}^{\infty} c_\nu z^\nu$, then if $|a_\nu| \leq M_1/\nu^h . r^\nu$, $|b_\nu| \leq M_2/\rho^\nu$, $1 \leq r < \rho$, we have $|c_\nu| \leq M/\nu^h . r^\nu$.
Thus we have

THEOREM 6.5.3. Under the hypotheses of Theorems 6.5.1 and 6.5.2 the conclusions are valid if the function f(z) is replaced by the product f(z)g(z) where g(z) is analytic in $|z| \leq 1$.

We also have in the above notation as a consequence of the inequalities

$$|a_\nu| \leq M_1/\nu^h . \rho^\nu, \quad |b_\nu| \leq M_2/\nu^l . \rho^\nu, \quad \rho \geq 1,$$

the following inequalities

$$|c_\nu| \leq \frac{M}{\nu^{h+l-1}\,\rho^\nu}, \quad 1 > h \geq l,$$

$$|c_\nu| \leqq \frac{M}{\nu^h \rho^\nu}, \; 1 \geqq h > 1, \; 1 > h \geqq 1,$$

$$|c_\nu| \leqq \frac{M \log \nu}{\nu^{h+1-1} \rho^\nu}, \; 1 = h \geqq 1 .$$

Thus we have the following

THEOREM 6.5.4. Let $f(z) = (z-z_1)^{h_1-1}$ $(z-z_2)^{h_2-1} = \sum_{\nu=0}^{\infty} a_\nu z^\nu$, $|z_1| = |z_2| = 1$, $z_1 \neq z_2$. Then we have

$$\left| f(z) - \sum_{\nu=0}^{n} a_\nu z^\nu \right| \leqq \epsilon_n, \; |z| \leqq 1/\rho,$$

where $\epsilon_n = M/n^{h_1 + h_2 - 1} \cdot \rho^n$, $1 > h_1 \geqq h_2$;

$\epsilon_n = M/n^{h_1} \cdot \rho^n$, $h_2 \geqq h_1 > 1$, $h_2 > h_1 \geqq 1$;

$\epsilon_n = M \log n/n^{h_1 + h_2 - 1} \cdot \rho^n$, $1 = h_1 \geqq h_2$.

The extension of Theorem 6.5.4 to functions of the form $\prod_{\nu=1}^{\lambda} (z-z_\nu)^{h_\nu - 1}$ is immediate; the details can be easily supplied by the reader.

The above theorems can be extended to approximation on an arbitrary analytic Jordan curve C by replacing the Taylor development by the Faber development of the function. It is to be observed that the degree of approximation which we obtain here follows from inequalities on the Taylor coefficients and from the fact that on $|z| = 1/\rho$ we have for the polynomial $p_n(z) = z^n$ on C the inequality $|p_n(z)| \leqq M/\rho^n$. For the Faber polynomial $F_n(z)$ we likewise have by (3.3.4) the inequality $|F_n(z)| \leqq M/\rho^n$ for z on C; we use here the Faber polynomials belonging to C_ρ to conform to the notation of Problem β. Furthermore by (3.3.8) we see that the Faber coefficient a_ν in the development $\sum a_\nu F_\nu(z)$ of a function f(z) is also the Taylor coefficient of a function $F(w) = -\sum_{\nu=0}^{\infty} a_\nu w^\nu$; the function F(w) is of precisely the same type as far as singularities and analyticity are concerned on $|w| = 1$ as the function f(z) on C_ρ, due to the analyticity of the curves, and of $\sum a_\nu \mathfrak{P}_\nu(1/w)$. Consequently we obtain inequalities on a_ν, the Faber coefficients, corresponding to the inequalities obtained above on the Taylor coefficients. The extension now follows by applying these inequalities on a_ν and $F_\nu(z)$ to $\sum_{\nu=n+1}^{\infty} a_\nu F_\nu(z)$ for z on C.

For C the segment $-1 \leqq z \leqq 1$ Bernstein [1926] studied the approximation of functions with isolated singularities on C_ρ; de la Vallée Poussin [1919] considered the analogous problem from the standpoint of trigonometric approximation; we shall return to these more special questions in Chapter VIII. Faber [1920] showed that his development yields the above degree of approximation to a function with a single pole of the first order on C_ρ; the extension to a function with a finite number of poles is mentioned by Sewell [1937]. The treatment in the present section follows closely that of Walsh and Sewell [1941].

§6.6. POLYNOMIALS OF BEST APPROXIMATION. If we denote by $L_n(z)$ the polynomial of best approximation to $f(z)$ on C in the sense of Tchebycheff the results of the preceding sections of this chapter can be collected to yield some interesting information on the degree of approximation of $L_n(z)$ to $f(z)$. Let

$$\max\ [|f(z) - L_n(z)|,\ z \text{ on } C] = \epsilon_n;$$

thus it follows that if $|f(z)-P_n(z)| \leqq \delta_n$, z on C, then $\epsilon_n \leqq \delta_n$.

We have, for example, the following

THEOREM 6.6.1. Let C be an analytic Jordan curve and let $f(z)$ belong to the class $L(k, \alpha)$ or $L'(k, \alpha)$ on C_ρ, $\rho > 1$. Then we have

$$(6.6.1)\qquad |f(z) - L_n(z)| \leqq \frac{M_1}{n^{k+\alpha}\rho^n},\quad z \text{ on } C,$$

(6.6.2) $$|f(z) - L_n(z)| \leqq \frac{M_2 r^n}{n^{k+\alpha} \rho^n},$$

z on C_r, $1 < r < \rho$,

and for $k \geqq 1$

(6.6.3) $$|f(z) - L_n(z)| \leqq \frac{M_3}{n^{k+\alpha-1}}, \quad z \text{ on } C_\rho .$$

More generally inequalities (6.6.2) and (6.6.3) follow in any case from inequality (6.6.1), whether $L_n(z)$ is the Tchebycheff polynomial of $f(z)$ or not.

Inequality (6.6.1) is an immediate consequence of Theorems 6.1.1 and 6.2.4. Inequality (6.6.2) follows by writing

$$f(z) = L_1(z) + [L_2(z) - L_1(z)] + \dots + [L_{n+1}(z) - L_n(z)] + \dots$$

and applying Theorem 2.1.3 to the inequality on the terms of this series resulting from inequality (6.6.1). Inequality (6.6.3) is obtained in a similar manner. The extensions of this theorem to more general boundaries in certain cases are immediate.

It is clear from inequality (6.6.2) that if $G_n(z)$ is the polynomial of best approximation to $f(z)$ on C_r, $1 < r < \rho$, in the sense of Tchebycheff and if we set $\max [|f(z) - G_n(z)|, z \text{ on } C_r] = \delta_n$, then $\delta_n \cdot n^{k+\alpha} \cdot \rho^n/r^n$ is bounded.

If $f(z)$ belongs to the class $L(k, \alpha)$ or $L'(k, \alpha)$ on C_ρ but to no higher class it is clear from Theorem 6.3.1 that $\epsilon_n \cdot n^{k+\alpha+1+\delta} \cdot \rho^n$, $\delta > 0$, is not bounded. We see from Theorem 6.5.1 that there exist functions of class $L(k, \alpha)$ and $L'(k, \alpha)$ for which $\epsilon_n \cdot n^{k+\alpha+1} \cdot \rho^n$ is bounded; on the other hand by §6.4 it is clear that there are functions of class $L(k, \alpha)$ and $L'(k, \alpha)$ for which $\epsilon_n \cdot n^{k+\alpha+\delta} \cdot \rho^n$, $\delta > 0$, is not bounded. Here, as below, we take C as an analytic Jordan curve.

For functions with isolated singularities we obtain a much better estimation of ϵ_n. For example let $f(z) = (z-z_o)^h$, z_o on C_ρ, where h is not a positive integer. Then by Theorem 6.5.1 we see that $\epsilon_n \cdot n^{h+1} \cdot \rho^n$ is bounded. On the other hand if we develop f(z) in the Faber polynomials $F_\nu(z)$ belonging to C, that is $f(z) = \sum_{\nu=0}^{\infty} a_\nu F_\nu(z)$, it follows from the results of §4.2 and the discussion of §6.5 that

$$(6.6.4) \qquad a_\nu = \frac{(1+\delta_\nu)}{\nu^{h+1}\rho^\nu}, \quad \lim_{\nu\to\infty} \delta_\nu = 0.$$

Now suppose $D_n(z) = f(z) - L_n(z)$ and develop $D_n(z)$ in a series of Faber polynomials

$$D_n(z) = \sum_{\nu=0}^{\infty} b_\nu^{(n)} F_\nu(z),$$

where $b_\nu^{(n)} = a_\nu$ for $\nu > n$ by virtue of the uniqueness of the Faber development. By (3.3.7) we have

$$|b_\nu^{(n)}| = \frac{1}{2\pi} \left| \int_{|w|=1} D_n(\psi(w)) \frac{dw}{w^{\nu+1}} \right| ;$$

hence $|b_\nu^{(n)}| \leqq 2\pi\epsilon_n$, $\nu = 1,2,\ldots$; consequently $|a_{n+1}| \leqq 2\pi\epsilon_n$ and hence by (6.6.4)

$$\epsilon_n \geqq \frac{M_1}{(n+1)^{h+1}\rho^{n+1}} \geqq \frac{M_2}{n^{h+1}\rho^n}.$$

Also if $f(z) = \log(z-z_o)$, z_o on C_ρ, we have

$$\frac{M_1}{n\rho^n} \leqq \epsilon_n \leqq \frac{M_2}{n\rho^n}.$$

The above discussion applies to more general functions of the type considered in §6.5.

The study of the degree of convergence on C_ρ and on $C_r (1 < r < \rho)$ (but not on E) of polynomials arising out of the study of Problem β has been called Problem γ by Walsh and Sewell [1940]; Theorem 6.6.1 is due to them. The estimation of ϵ_n is due to Faber [1920] for a function f(z) with a single pole of the first order, and the extension to several poles to Sewell [1937]. Best approximation to functions with algebraic singularities is considered by Walsh and Sewell [1940, 1941].

§6.7. EXERCISES. 6.7.1. Let E be a closed limited set bounded by a contour C, and let C_ρ also be a contour. Let the function f(z) be of class L(k, α), k ≥ 0, on C_ρ. Let $p_n(z)$ denote the unique polynomial which interpolates to f(z) in n+1 equally distributed points on C. Then we have

$$|f(z) - p_n(z)| \leq \frac{M}{n^{k+\alpha} \rho^n}, \quad z \text{ on } E.$$

[Walsh and Sewell, 1940]

6.7.2. Let p(z) be the polynomial $p(z) \equiv$ $\equiv (z-\beta_1) \dots (z-\beta_\lambda)$, and let Γ_μ denote generically the lemniscate $|p(z)| = \mu > 0$. Let Γ_{μ_1} consist of a finite number of mutually exterior analytic Jordan curves. Let f(z) be of class L(k, α), k ≥ 0, on Γ_{μ_1}. Let $p_n(z)$ denote the unique polynomial which interpolates to f(z) in the n+1 ($= q\lambda + r$, $0 \leq r < \lambda$) points namely $\beta_1, \beta_2, \dots, \beta_r$ counted each q+1 times and the points $\beta_{r+1}, \dots, \beta_\lambda$ counted each q times. Then we have for each $\mu (0 < \mu < \mu_1)$

$$|f(z) - p_n(z)| \leq \frac{M}{n^{k+\alpha}} \left[\frac{\mu}{\mu_1}\right]^{n/\lambda}, \quad z \text{ on } \Gamma_\mu.$$

[Walsh and Sewell, 1940]

6.7.3. Let C be an analytic Jordan curve and let f(z) be of class L(k, α) on C_ρ. Let $p_n(z)$ denote the unique polynomial which interpolates to f(z) in the n+1 roots of either (i) $T_{n+1}(z)$, the Tchebycheff polynomial belonging to C; (ii) $F_{n+1}(z)$, the Faber polynomial belonging to C; or (iii) $Q_{n+1}(z)$, the polynomial of degree n+1 of the set of polynomials normal and orthogonal on C with respect to a positive analytic weight function. Then we have for n sufficiently large

$$(6.7.1) \quad |f(z) - p_n(z)| \leq \frac{M}{n^{k+\alpha}\rho^n}, \quad z \text{ on } E.$$

[Walsh and Sewell, 1940, for $k \geq 0$]

NOTE 1. The Tchebycheff polynomial $T_n(z)$ belonging to C is the uniquely determined polynomial of degree n of the form

$$T_n(z) = z^n + a_1 z^{n-1} + \dots + a_n,$$

for which

$$\max\,[|T_n(z)|, \; z \text{ on } C]$$

is least. Faber [1920] has shown that

$$\lim_{n \to \infty} \frac{T_n(z)}{e^{ng}[\phi(z)]^n} = 1,$$

uniformly for z on or exterior to E.

NOTE 2. It is true that for small n some roots of the Tchebycheff, Faber, and orthogonal polynomials of degree n belonging to C may lie on or exterior to a given C_ρ ; but when C_ρ is given, all roots of such polynomials for n sufficiently large lie interior to C_ρ . For the validity of the usual proof of (5.1.1), however, it is necessary that all the points of interpolation lie interior to C_ρ . Consequently if $f(z)$ is given analytic interior to a particular C_ρ , the polynomials $p_n(z)$ of interpolation in the roots of the Tchebycheff, Faber, and orthogonal polynomials need not be defined when n is small, but surely are defined for n sufficiently large. This restriction on n is ordinarily nor serious, however, for we may define $f(z)$ artificially in any way at the points of interpolation on or exterior to C_ρ ; such an inequality as (6.7.1) above, if valid for n sufficiently large, remains valid for all n provided the constant M is suitably altered.

Furthermore if $P_n(z)$ represents the Faber,

Tchebycheff, or orthogonal polynomial of degree n we have $|P_n(z)/P_n(t)| \leq M/\rho^n$, for z on C and t on C_ρ, for n sufficiently large, but not necessarily for all n.

6.7.4. Case (ii) of Ex. 6.7.3 for C a curve of the type described in Ex. 5.5.3. [Walsh and Sewell, 1940, for $k \geq 0$]

6.7.5. In Case (iii) of Ex. 6.7.3 we have

$$|f(z) - p_n(z)| \leq \frac{M_1 \sigma^n}{n^{k+\alpha} \rho^n}, \quad z \text{ on } C_\sigma,$$

for $\sigma < 1$ but $1-\sigma$ sufficiently small. [Walsh and Sewell, 1940, for $k \geq 0$]

6.7.6. Let C be a contour and let f(z) be analytic interior to C_ρ and of class F_q on C_ρ. Let $p_n(z)$ interpolate to f(z) in n+1 equally distributed points on C. Then we have

$$|f(z) - p_n(z)| \leq \frac{M}{n^{q-1/2}\rho^n}, \quad z \text{ on } C.$$

[Walsh and Sewell, 1940]

6.7.7. Let C and C_ρ be contours and let f(z) be of class $L(k, \alpha)$, $k \geq 1$, on C_ρ. Let $p_n(z)$ be the polynomial which interpolates to f(z) in n+1 equally distributed points on C. Then we have

$$|f(z) - p_n(z)| \leq M/n^{k+\alpha-1}, \quad z \text{ on } C_\rho.$$

[Walsh and Sewell, 1940]

6.7.8. Let C be an analytic Jordan curve and let f(z) be of class $L(k, \alpha)$, $k \geq 1$, on C_ρ. Let $p_n(z)$ denote the polynomial which interpolates to f(z) in the n+1 roots of $T_{n+1}(z)$, $F_{n+1}(z)$, or $Q_{n+1}(z)$. Then we have

$$|f(z) - p_n(z)| \leq M/n^{k+\alpha-1}, \quad z \text{ on } C_\rho.$$

[Walsh and Sewell, 1940]

6.7.9. Let E be the closed interior of a Jordan curve C which is a contour and let f(z) belong to the class F_q on C_ρ. Let $p_n(z)$ interpolate to f(z) in n+1 equally distributed points on C. Then we have

$$|f(z) - p_n(z)| \leqq \frac{M}{n^{q-1/2}\rho^n}, \quad z \text{ on } C.$$

[Walsh and Sewell, 1940]

6.7.10. Let f(z) belong to the class L(k, α) or L'(k, α) on $|z| = 1$, and let $f(z) = \sum_{\nu=0}^{\infty} a_\nu z^\nu$, $|z| < 1$, then

$$\left|f(z) - \sum_{\nu=0}^{n} a_\nu z^\nu\right| \leqq \frac{M}{n^{k+\alpha}\rho^n}, \quad |z| \leqq 1/\rho .$$

[Walsh and Sewell, 1941]

6.7.11. Let f(z) be analytic in $|z| < 1$ and continuous on $|z| \leqq 1$, and let $P_n(z)$ exist such that $|f(z)-P_n(z)| \leqq \epsilon_n$, $|z| = 1$. Then we have

$$\left|f(z) - \sum_{\nu=0}^{n} a_\nu z^\nu\right| \leqq M\epsilon_n/\rho^n, \quad |z| \leqq 1/\rho .$$

[Walsh and Sewell 1941]

6.7.12. Under the hypothesis of Ex. 6.7.11 we have

$$\left|\int_0^z \left|f(z) - \sum_{\nu=0}^{n} a_\nu z^\nu\right| dz\right| \leqq \frac{M\epsilon_n}{n\rho^n}, \quad |z| \leqq 1/\rho .$$

[Walsh and Sewell, 1941]

6.7.13. Under the hypothesis of Ex. 6.7.11 we have

$$\left|f'(z) - \sum_{\nu=1}^{n} \nu a_\nu z^{\nu-1}\right| \leqq Mn\epsilon_n/\rho^n.$$

[Walsh and Sewell, 1941]

6.7.14. Let $f(z) = \sum^{\infty} a_\nu z^\nu$, belong to the class L(k, α) on $|z| = 1$. Then we have

$$\left|f'(z) - \sum_{\nu=1}^{n} \nu a_\nu z^{\nu-1}\right| \leqq \frac{M}{n^{k+\alpha-1}\rho^n}, \quad |z| \leqq 1/\rho .$$

[Walsh and Sewell, 1941]

6.7.15. Let C be an analytic Jordan curve and let f(z) be analytic interior to C and continuous on $\overline{C}$. Let $P_n(z)$ exist such that

$$|f(z) - P_n(z)| \leqq \epsilon_n, \quad z \text{ on } C_\rho .$$

Then we have

$$\left| f(z) - \sum_{\nu=0}^{n} a_\nu F_\nu(z) \right| \leqq M \epsilon_n / \rho^n, \ z \text{ on } C,$$

where $\sum a_\nu F_\nu(z)$ is the development of $f(z)$ in the Faber polynomials belonging to C. [Walsh and Sewell, 1941]

6.7.16. Let C be a Jordan curve which is a contour and let $f(z)$ be analytic in C_ρ and continuous on $\overline{C}_\rho$. Let $f(z) \sim \sum_{\nu=0}^{\infty} a_\nu Q_\nu(z)$, $a_\nu = \int_{C_\rho} f(z) \overline{Q}_\nu(z) |dz|$, where $Q_\nu(z)$ belongs to the set of polynomials normal and orthogonal on C_ρ. Let $p_n(z)$ be the polynomial which interpolates to $f(z)$ in n+1 equally distributed points on C. Then we have

$$\lim_{n \to \infty} \rho^n \max [|f(z) - p_n(z)|, \ z \text{ on } C] = 0.$$

More generally this conclusion holds if $f(z)$ belongs to the class H_2[see, e.g., Walsh, 1935] on C_ρ. [Walsh and Sewell]

Note: Let $f(z)$ be analytic in $|z| < 1$ and have boundary values almost everywhere on $|z| = 1$. We say [Walsh and Sewell, 1941] that $f(z)$ belongs to the class $L_m(0, \alpha)$, $0 < \alpha \leqq 1$, $m > 0$, provided the integrated Lipschitz condition is satisfied

$$(6.7.1) \quad \left[\frac{1}{2\pi} \int_{-\pi}^{\pi} |f(e^{i\theta + ih}) - f(e^{i\theta})|^m d\theta \right]^{1/m} \leqq Lh^\alpha,$$

where L is a constant independent of h; we say that $f(z)$ belongs to the class $L_m(k, \alpha)$ where k is a negative integer provided we have

$$(6.7.2) \quad \left[\frac{1}{2\pi} \int_{-\pi}^{\pi} |f(re^{i\theta})|^m d\theta \right]^{1/m} \leqq L(1-r)^{k+\alpha}, \ r < 1,$$

$0 < \alpha \leqq 1$,
where L is a constant independent of r.

Hardy and Littlewood [1932] have proved: (i) if $k + \alpha \leqq 0$ and if $f(z)$ is of class $L_m(k, \alpha)$, then $f^{(j)}(z)$ is of class $L_m(k-j, \alpha)$; (ii) if $k+\alpha < 0$, $k+\alpha-q < 0$, and if $f(z)$ is of class $L_m(k-q, \alpha)$, then the q-th integral of $f(z)$ is of class $L_m(k, \alpha)$; (iii) a necessary and sufficient condition that $f(z)$ be of class $L_m(0, \alpha)$. $0 < \alpha \leq 1$, is that $f'(z)$ be of class $L_m(-1, \alpha)$.

We now say [Walsh and Sewell, 1941] that $f(z)$ is of class $L_m(k, \alpha)$, where k is a non-negative integer and $0 < \alpha \leq 1$ provided $f^{(k+1)}(z)$ is of class $L_m(-1, \alpha)$. We define the class $L'_m(-1,1)$ as the class of integrals of functions of class $L_m(-2,1)$, and the class $L'_m(k,1)$, $k>-2$, as the class of $(k+2)$-th iterated integrals of functions of class $L'_m(-2,1)$.

6.7.17. If $f(z)$ is of class $L_m(k, \alpha)$, then the function $f^{(j)}(z)$ is of class $L_m(k-j, \alpha)$, and if $(k+\alpha)(k+j+\alpha+1)$ is not a negative integer the j-th integral of $f(z)$ is of class $L_m(k+j, \alpha)$. [Walsh and Sewell, 1941]

6.7.18. If $f(z)$ is of class $L'_m(k,1)$, $k > -2$, then $f^{(j)}(z)$, $0 \leq j \leq k+2$, is of class $L'_m(k-j,1)$; also $f^{(j)}(z)$, $j > k+2$ is of class $L'_m(k-j,1)$. If $f(z)$ is of class $L_m(k,1)$, $k < -2$, then $f^{(j)}(z)$ is of class $L_m(k-j, \alpha)$ and the j-th iterated integral of $f(z)$ is of class $L_m(k+j,1)$ or $L'_m(k+j,1)$ according as $k+j \leq -2$ or $k+j > -2$. [Walsh and Sewell, 1941]

6.7.19. If $f(z)$ is of class $L_2(k, \alpha)$ or of class $L'_2(k, \alpha)$, there exist polynomials $p_n(z)$ such that we have for $|z| = 1/\rho < 1$

$$(6.7.3) \quad |f(z) - p_n(z)| \leq \frac{M}{n^{k+\alpha} \rho^n}.$$

[Walsh and Sewell, 1941]

6.7.20. Let $f(z)$ be defined in $|z| < 1$ and let $p_n(z)$ exist such that

$$(6.7.4) \quad |f(z) - p_n(z)| \leq \frac{M}{n^{k+\alpha+1/2} \rho^n}$$

is valid for $|z| \leq 1/\rho < 1$; then $f(z)$ is of class $L_2(k, \alpha)$ if $k + \alpha + 1$ is not a positive integer and is of class $L'_2(k, \alpha)$ if $k + \alpha + 1$ is a positive integer.

[Walsh and Sewell, 1941]

6.7.21. In Ex. 6.7.20 the hypothesis may be taken as

$$\int_{|z|=1/\rho} |f(z) - p_n(z)|^m |dz| \leqq \frac{M}{n^{m(k+\alpha+1/2)}\rho^{mn}}, \quad m > 0.$$

[Walsh and Sewell, 1941]

6.7.22. Show that in Exs. 6.7.19 and 6.7.20 we have in a sense the best possible results, namely in the sense that we cannot replace $k+\alpha$ in (6.7.3) by any $\alpha' > k + \alpha$, and that we cannot replace $k + \alpha$ in (6.7.4) by any $\alpha'' < k + \alpha$. [Walsh and Sewell, 1941]

6.7.23. Let $f(z) = \sum_{\nu=0}^{\infty} a_\nu z^\nu$ be analytic in $|z| < 1$ and let $\sum_{\nu=0}^{\infty} |a_\nu|^2$ converge. Then $p_n(z)$ exists such that

$$|f(z) - p_n(z)| \leqq \delta(n)/\rho^n, \quad |z| \leqq 1/\rho < 1,$$

where $\delta(n)$ approaches 0 as n becomes infinite. [Walsh]

§6.8. DISCUSSION. As a consequence of Ex. 6.7.2 we see that if $f(z)$ ($= \sum_{\nu=0}^{\infty} a_\nu z^\nu$) belongs to the class $L(k, \alpha)$, $k \geq 0$, on $|z| = 1$, then we have $\left|f(z) - \sum_{\nu=0}^{n} a_\nu z^\nu\right| \leqq M/n^{k+\alpha} \cdot \rho^n$, $|z| \leqq 1/\rho < 1$; whereas on $|z| = 1$ we have (Ex. 3.7.2) only $\left|f(z) - \sum_{\nu=0}^{n} a_\nu z^\nu\right| \leqq M \log n/n^{k+\alpha}$. Of course we know that $\sum_{\nu=0}^{n} a_\nu z^\nu$ is not the polynomial of degree n of best approximation to $f(z)$ on $|z| = 1$, but the above observation raises some questions as far as $|z| = 1/\rho$ is concerned. Let $f(z)$ be analytic in $|z| < 1$; let $p_n(z)$ be the sum of the first n+1 terms of the Taylor development of $f(z)$ and let $L_n(z)$ be the polynomial of degree n of best approximation to $f(z)$ on $|z| = 1/\rho < 1$ in the sense of Tchebycheff. Let us set

$$\epsilon_n = \max\,[\,|f(z)-p_n(z)|,\ |z| = 1/\rho\,],$$

$$\delta_n = \max\,[\,|f(z)-L_n(z)|,\ |z| = 1/\rho\,].$$

Is it true that ϵ_n/δ_n is bounded? Is it true that if q exists such that with q' > q the quantity ϵ_n satisfies no inequality of the form

$$\epsilon_n \leqq \frac{M}{n^{q'} \rho^n}, \quad q' > q,$$

then δ_n can satisfy no inequality of the form

$$\delta_n < \frac{M_1}{n^{q'} \rho^n}, \quad q' > q \quad ?$$

We remark in this connection that $p_n(z)$ interpolates to f(z) in the origin counted n+1 times, that $p_n'(z)$ interpolates to f'(z) in the origin counted n times, and that the integral of $p_n(z)$ interpolates to the integral of f(z), where the constants of integration are suitably chosen, in the origin counted n+2 times; thus, in a sense, the polynomial $p_n(z)$ is invariant under differentiation and integration. Do other polynomials exist which are in this same sense invariant under differentiation and integration? Under what conditions on C and on the points of interpolation is it true that if $p_n(z)$ interpolates to f(z) in n+1 points, then $p_n'(z)$ interpolates to f'(z) in n points? Under what conditions on C and on the points of interpolation is it true that if $p_n(z)$ interpolates to f(z) in n+1 points, then the integral of $p_n(z)$ interpolates to the integral of f(z) in n+2 points, where the constants of integration are suitably chosen?

In addition to the exceptional case of $\alpha = 1$ in Problem α which forced us to introduce the class Log(k,1) we find ourselves confronted with new complications here and consequently a new class L'(k,1). Of course Theorem 5.2.5 sheds some light on the relation between these two classes, but a thorough investigation should yield new and interesting results.

There is always the problem of extending the results to sets bounded by more general curves. To mention one example, Theorem 6.2.3 is very important in our theory and the proof given here is entirely dependent upon the analyticity of the curve; an extension of this theorem to more general curves would be an important contribution. Also there is the obvious problem of connecting the class $L(k, \alpha)$ for $k \geq 0$ to the class $L(k, \alpha)$ for $k < 0$, where C is a contour.

As in Problem α there is the problem of studying degree of approximation on lemniscates and other sets for which C_ρ has multiple points.

The classes $L(k, \alpha)$ and $L'(k, \alpha)$ are the natural classes to consider in studying degree of approximation on sets for which C_ρ consists of an analytic Jordan curve. What are the natural classes to consider in the case of lemniscates and other sets for which C_ρ has multiple points, classes which will lead to the degrees of approximation of the form M/n^q, and of the form $M/n^q \cdot \rho^n$?

It is to be observed that in Theorem 6.1.1 if C is a contour and if $k < 0$ then the inequality

$$|f(z)| \leq L(\rho - r)^{k+\alpha}, \quad z \text{ on } C_r,$$

where L is a constant independent of r, leads to $p_n(z)$ satisfying inequality (6.1.1), even if C_ρ has multiple points. Also in Theorem 6.3.1 if C is a contour and $k < 0$, we have

$$|f(z)| \leq L(\rho - r)^{k+\alpha}, \quad z \text{ on } C_r, \quad k+\alpha < 0,$$

or

$$|f(z)| \leq L|\log(\rho - r)|, z \text{ on } C_r, \quad k+\alpha = 0,$$

where L is a constant independent of r, even if C_ρ has multiple points. Such functions deserve further study.

In connection with functions of class $L_m(k, \alpha)$ (Exs. 6.7.17-6.7.22) there are many open problems. It is true [Walsh and Sewell, 1941] that Ex. 6.7.19 extends to the case that C is an analytic Jordan curve, but

Ex. 6.7.20 has not as yet been extended. Both Exs. 6.7.19 and 6.7.20 are stated for $m = 2$; we have the obvious problem of extending these results to include an arbitrary $m > 0$. It may be observed [Walsh and Sewell, 1941] that for $m > 1$ the Hölder inequality (4.5.1) and for $m = 1$ more elementary methods establish Ex. 6.7.19, where now $f(z)$ is a function of class $L_m(k, \alpha)$ or of class $L_m'(k, \alpha)$. It is to be observed that functions of the class $L_m(-1, 1)$ are identical with functions of the class H_δ studied by F. Riesz [1923].

CHAPTER VII

APPROXIMATION MEASURED BY A LINE INTEGRAL

§7.1. DIRECT THEOREMS. Just as in Problem α, for approximation in the sense of weighted p-th powers the direct theorems of Problem β are immediate consequences of the direct Tchebycheff theorems. Thus from the results of §§6.1 and 6.2 we obtain immediately several results.

From Theorem 6.1.1 we have

THEOREM 7.1.1. Let C be a contour and let C_ρ also be a contour. Let f(z) belong to the class $L(k, \alpha)$ on C_ρ. Then $p_n(z)$ exists such that

$$(7.1.1) \quad \int_C \Delta(z)|f(z)-p_n(z)|^p \, |dz| \leqq \frac{M}{n^{(k+\alpha)p} \rho^{np}}, \quad p > 0,$$

where $\Delta(z)$ is positive and continuous on C.

From Theorem 6.2.4 we have

THEOREM 7.1.2. Let C be an analytic Jordan curve and let f(z) belong to the class $L'(k, \alpha)$ on C_ρ. Then $p_n(z)$ exists such that (7.1.1) is valid.

From Theorem 6.1.2 we have the following extension of Theorem 7.1.1:

THEOREM 7.1.3. The conclusion of Theorem 7.1.1 is valid for $k \geq 0$ if C is a lemniscate provided C_ρ is a contour.

It is clear from the discussion of §6.4 concerning the direct Tchebycheff theorems that the above results are the best possible in the usual sense, that is to say, the exponent of n in the right hand member of inequality (7.1.1) cannot be increased in general. Our example is for the case $p = 2$, C the circle, and $\Delta(z) \equiv 1$.

Further direct theorems for specific sequences are found in the exercises.

Theorem 7.1.1 is due to Walsh and Sewell [1941] and Theorem 7.1.2 to Walsh and Sewell [1940].

§7.2. INDIRECT THEOREMS. The indirect theorems corresponding to the direct theorems of the preceding section are easily proved thanks to the general inequalities (4.2.5) and to Theorems 5.3.1 and 5.3.2 which enable us to obtain a Tchebycheff degree of approximation exterior to C. Then we have merely to call on the theorems of §6.3 to obtain the final results. Of course the discussion of §4.2 concerning the definition of $f(z)$ on C or almost everywhere on C is also appropriate here.

Our first theorem is

THEOREM 7.2.1. Let C be a rectifiable Jordan curve or arc and let $f(z)$ be defined on C; let $P_n(z)$ exist such that

$$(7.2.1) \quad \int_C \Delta(z)|f(z)-P_n(z)|^p\,|dz| \leq \frac{M}{n^{(k+\alpha+1)p}\,\rho^{np}}, \quad p > 0,$$

where $\Delta(z)$ is positive and continuous on C. Then $f(z) = f_1(z)$ almost everywhere on C, where

$f_1(z) = \lim_{n\to\infty} P_n(z)$, z interior to C_ρ ;

furthermore $f_1(z)$ belongs to the class $L(k, \alpha)$ on C_ρ if $k+\alpha+1$ is not a positive integer and to the class $L'(k, \alpha)$ if $k + \alpha + 1$ is a positive integer.

As in the proof of Theorem 4.2.1 we obtain from inequality (7.2.1) by virtue of the boundedness on C of $1/\Delta(z)$ and the general inequalities (4.2.5)

$$\int_C |P_{n+1}(z) - P_n(z)|^p \, |dz| \leq \frac{2M}{n^{(k+\alpha+1)p}\rho^{np}} .$$

Choose r_0, $1 < r_0 < \rho$; then by Theorem 5.3.1 we have

$$(7.2.2) \quad |P_{n+1}(z) - P_n(z)| \leq \frac{M_1 r_0^n}{n^{k+\alpha+1}\rho^n}, \quad z \text{ on } C_{r_0} ,$$

where M_1 depends on r_0. Now if we write

$$f_1(z) = P_1(z) + [P_2(z) - P_1(z)] + \dots$$
$$+[P_{n+1}(z) - P_n(z)] + \dots$$

there follows from inequality (7.2.2) by the usual argument

$$(7.2.3) \quad |f_1(z) - P_n(z)| \leq \frac{M_2}{n^{k+\alpha+1}(\rho/r_0)^n}, \quad z \text{ on } C_{r_0} .$$

Since inequality (7.2.3) is valid for every n we have merely to recall (§2.3) that if we denote C_{r_0} by Γ then the curve C_ρ coincides with Γ_{ρ/r_0} and to apply Theorem 6.3.1. The equality of $f(z)$ and $f_1(z)$ almost everywhere on C follows by the argument of §4.2.

By virtue of Theorem 5.3.2. along with the subsequent remarks and Theorem 6.3.2. we have

THEOREM 7.2.2. Let C be a contour or a lemniscate and let $f(z)$ be defined on C; let $P_n(z)$ exist such that inequality (7.2.1) is valid for $k \geq 0$. Then $f(z) = f_1(z)$ almost everywhere on C, where $f_1(z) = \lim_{n \to \infty} P_n(z)$, z on $\overline{C}_\rho$; furthermore $f_1(z)$ belongs to the class $L(k, \alpha)$ on C_ρ if $\alpha \neq 1$ and to the class $\mathrm{Log}(k,1)$ on C_ρ if $\alpha = 1$.

The examples of §§6.3 and 6.4 used in connection with the indirect Tchebycheff theorems serve to show that the above results are the best possible in the sense previously described ($p = 2$); the details are left to the reader.

In the proof of Theorem 7.2.1 it is to be observed that inequality (7.2.3) yields information on the degree of convergence of the sequence $p_n(z)$ on C_r, $1 < r_0 < r < \rho$, and hence further results on Problem γ (compare §6.6).

The conclusion of Theorem 7.2.1 is mentioned by Walsh and Sewell [1941] for C an analytic Jordan curve. Theorem 7.2.2 for a single Jordan curve is due to Sewell [1939a] and as stated here to Walsh and Sewell [1940].

§7.3. ORTHOGONAL POLYNOMIALS. In Problem β there are two possibilities as far as orthogonal polynomials are concerned since we deal with two sets of curves. We begin by considering polynomials normal and orthogonal on C, rather than on C_ρ, and establish theorems analogous to those of §4.3. For simplicity we take $\Delta(z) \equiv 1$.

By virtue of Theorem 7.1.1 and the least square property of the orthogonal development we have

THEOREM 7.3.1. Let C be an analytic Jordan curve and let $Q_\nu(z)$, $\nu = 0,1,2,\ldots$, be the set of polynomials normal and orthogonal on C. Let $f(z)$ belong to the class $L(k, \alpha)$ or $L'(k, \alpha)$ on C_ρ, and let

$$(7.3.1) \quad a_\nu = \int_C f(z)\, \overline{Q_\nu(z)}\, |dz|.$$

Then we have

$$(7.3.2)\quad \int_C \left| f(z) - \sum_{\nu=0}^{\infty} a_\nu \, Q_\nu \, (z) \right|^2 |dz| = \sum_{\nu=n+1}^{\infty} |a_\nu|^2 \leqq \frac{M}{n^{(k+\alpha)2} \quad \rho^{2n}}.$$

Inequality (7.3.2) yields

COROLLARY 7.3.2. Under the hypothesis of Theorem 7.3.1 we have

$$(7.3.3)\quad |a_n| \leqq \frac{M}{n^{k+\alpha} \quad \rho^n} \cdot$$

By virtue of the boundedness of $Q_\nu\,(z)$ on C from formula (4.3.6) and by inequality (7.3.3) we have

COROLLARY 7.3.3. Under the hypothesis of Theorem 7.3.1 we have

$$\left| f(z) - \sum_{\nu=0}^{n} a_\nu \, Q_\nu \, (z) \right| \leqq \frac{M}{n^{k+\alpha} \quad \rho^n}, \quad z \text{ on } C.$$

The method used in deriving inequality (7.2.3) in the proof of Theorem 7.2.1 yields

COROLLARY 7.3.4. Under the hypothesis of Theorem 7.3.1 we have

$$\left| f(z) - \sum_{\nu=0}^{n} a_\nu \, Q_\nu \, (z) \right| \leqq \frac{M\, r^n}{n^{k+\alpha} \quad \rho^n}, \quad z \text{ on } C_r,$$

$1 < r < \rho$,

where M depends on r.

Another corollary which follows from (4.3.6) and inequality (7.3.3) is

COROLLARY 7.3.5. Under the hypothesis of

Theorem 7.3.1 we have for $k \geq 1$

$$\left| f(z) - \sum_{\nu=0}^{n} a_\nu Q_\nu(z) \right| \leq \frac{M}{n^{k+\alpha-1}}, \quad z \text{ on } C_\rho .$$

Let C be an analytic Jordan curve and let us consider polynomials $Q_\nu(z)$ which are normal and orthogonal on C_ρ. Suppose f(z) belongs to the class (see §4.3) F_q on C_ρ, so that we have $|a_\nu| \leq M/\nu^q$, where

$$a_\nu = \int_{C_\rho} f(z) \overline{Q_\nu(z)} \, |dz|$$

Then by virtue of the fact that the formula (4.3.6) is valid for z on C we have

THEOREM 7.3.6. Let C be an analytic Jordan curve and let $Q_\nu(z)$, $\nu = 0, 1, 2, \ldots,$ be the set of polynomials normal and orthogonal on C_ρ. Let f(z) belong to the class F_q on C_ρ and let

$$a_\nu = \int_{C_\rho} f(z) \overline{Q_\nu(z)} \, |dz| .$$

Then we have

$$\left| f(z) - \sum_{\nu=0}^{n} a_\nu Q_\nu(z) \right| \leq \frac{M}{n^q \rho^n}, \quad z \text{ on } \overline{C}.$$

This theorem is interesting, particularly in comparison with Ex. 6.7.8.

§7.4. POLYNOMIALS OF BEST APPROXIMATION. In summarizing our results on approximation in the sense of least weighted p-th powers we have the following theorem:

THEOREM 7.4.1. Let C be an analytic Jordan curve and let f(z) belong to the class $L(k, \alpha)$ or $L'(k, \alpha)$ on C_ρ. Let $K_n(z)$ denote a polynomial of best approximation to f(z) on C in the sense of least weighted p-th powers,

$p > 0$, with the weight function $\Delta(z)$ positive and continuous on C. Then we have

$$(7.4.1) \quad \int_C \Delta(z)|f(z)-K_n(z)|^p |dz| \leq \frac{M_1}{n^{(k+\alpha)p}\rho^{np}},$$

$$(7.4.2) \quad |f(z)-K_n(z)| \leq \frac{M_2}{n^{k+\alpha-1/p}\rho^n}, \quad z \text{ on } C,$$

$$(7.4.3) \quad |f(z)-K_n(z)| \leq \frac{M_3 r^n}{n^{k+\alpha}\rho^n}, \quad z \text{ on } C_r, 1 < r < \rho,$$

$$(7.4.4) \quad |f(z)-K_n(z)| \leq \frac{M_4}{n^{k+\alpha-1}}, \quad z \text{ on } C_\rho,$$

where the validity of (7.4.4) requires $k \geq 1$.

Inequality (7.4.1) is an immediate consequence of Theorem 7.1.1 and the definition of $K_n(z)$. Inequality (7.4.2) follows from Theorems 6.1.1 or 6.2.4 along with Theorem 4.4.1. Inequalities (7.4.3) and (7.4.4) follow from inequality (7.4.1) by the method used in the proof of Theorem 7.2.1.

By applying Theorem 7.1.2 instead of Theorem 7.1.1 we have

THEOREM 7.4.2. In Theorem 7.4.1 let C be a contour or a lemniscate and let C_ρ be a contour. Let f(z) belong to the class $L(k, \alpha)$, $k \geq 0$, on C_ρ. Then inequalities (7.4.1) and (7.4.3) are valid, and inequality (7.4.4) holds provided $k \geq 1$.

We leave to the reader the discussion of polynomials of best approximation to f(z) in C_r, $1 < r < \rho$, (Problem γ) in the sense of least weighted p-th powers.

Theorem 7.4.2 is due to Walsh and Sewell [1940].

§7.5. GENERALITY OF THE WEIGHT FUNCTION. As in

Problem α it is not in general necessary to require that the weight function $\Delta(z)$ be positive and continuous on C. In fact for the direct theorems of §7.1 the argument used in the same connection in §4.5 shows that we obtain the same conclusions provided the weight function $\Delta(z)$ is non-negative and integrable Lebesgue on C.

For the indirect theorems of Problem β we can use without loss a weight function which satisfies only mild restrictions, whereas the use of these same restrictions on the weight function in Problem α is followed by a weakening of our conclusions. Suppose in Theorem 7.2.1 for example that the non-negative weight function $\Delta(z)$ is integrable and that some negative power of $\Delta(z)$ is integrable. Then by the Hölder inequality (4.5.1) we have as in §4.5 and in the same notation

$$\int \left| f(z) - P_n(z) \right|^{p(1-\delta)} |dz|$$

$$\leqq \left[\int \frac{|dz|}{[\Delta(z)]^{(1-\delta)/\delta}} \right]^{\delta} \left[\int \Delta(z) \left| f(z) - P_n(z) \right|^p |dz| \right]^{1-\delta}, \quad 0 < \delta < 1.$$

Thus from inequality (7.2.1) we obtain

$$\int_C \left| f(z) - P_n(z) \right|^{p(1-\delta)} |dz|$$

$$\leqq \frac{M}{n^{(k+\alpha+1)p(1-\delta)} \rho^{np(1-\delta)}} \quad ;$$

we now have merely to replace p by $p(1-\delta)$ to obtain the same conclusions by the same methods. It is clear, for example, that inequality (7.2.2) follows in this case just as in the original proof.

§7.6. EXERCISES. 7.6.1. Let C be an analytic Jordan curve and let $f(z)$ belong to the class $L(k,\alpha)$. Let $p_n(z)$ be the polynomial of degree n which interpolates to $f(z)$ in $n + 1$ equally distributed points on C. Then we have

$$(7.6.1)\quad \int_C \Delta(z)\left|f(z) - p_n(z)\right|^p |dz| \leq \frac{M}{n^{(k+\alpha)p}\rho^{np}},\ p > 0,$$

where $\Delta(z)$ is positive and continuous on C. [Walsh and Sewell, 1940]

7.6.2. In Ex. 7.6.1 the equally distributed points on C may be replaced by the roots of the Tchebycheff, Faber, or orthogonal polynomials belonging to C. [Walsh and Sewell, 1940]

7.6.3. The conclusion of Ex. 7.6.1 is valid for C a contour or a lemniscate for which C_ρ consists of mutually exterior Jordan curves, provided $k \geq 0$. [Walsh and Sewell, 1940]

7.6.4. In Ex. 7.6.1 we have

$$(7.6.2)\quad \iint_{\overline{C}} \Delta(z)\left|f(z) - p_n(z)\right|^p dS \leq \frac{M}{n^{(k+\alpha)p}\rho^{np}},\ p > 0.$$

7.6.5. Under the hypothesis of Ex. 7.6.2 we have inequality (7.6.2).

7.6.6. Under the hypothesis of Ex. 7.6.3 we have inequality (7.6.2).

7.6.7. Let $f(z)\left(= \sum_{\nu=0}^{\infty} a_\nu z^\nu\right)$ belong to the class $L(k, \alpha)$ on $|z| = 1$. Then

$$\iint_{|z| \leq 1/\rho < 1} \left|f(z) - \sum_{\nu=0}^{n} a_\nu z^\nu\right|^2 dS \leq \frac{M}{n^{2(k+\alpha)+1}\rho^{2n}}.$$

7.6.8. Let C be a contour or a lemniscate and let C_ρ consist of mutually exterior curves. Let $R_\nu(z)$, $\nu = 0,1,2,\ldots$, be the set of polynomials normal and orthogonal on $\overline{C}$. Let $f(z)$ belong to the class $L(k, \alpha)$ on C_ρ, $k \geq 0$, and let $b_\nu = \iint_{\overline{C}} f(z)\overline{R}_\nu(z)dS$. Then we have

$$\iint_{\overline{C}} \left|f(z) - \sum_{\nu=0}^{n} b_\nu R_\nu(z)\right|^2 dS \leq \frac{M}{n^{(k+\alpha)2}\rho^{2n}}.$$

7.6.9. In Ex. 7.6.8 we have $|b_\nu| \leq M/n^{k+\alpha}\cdot\rho^n$.

7.6.10. Let C be an analytic Jordan curve; then the conclusions of Ex. 7.6.8 and Ex. 7.6.9 are valid for all k.

7.6.11. In Ex. 7.6.8 let C be a Jordan curve which is a contour; then for $k \geqq 1$ we have

$$\left| f(z) - \sum_{\nu=0}^{n} b_\nu R_\nu(z) \right| \leqq \frac{M}{n^{k+\alpha-1} \rho^n}, \quad z \text{ on } C.$$

7.6.12. In Ex. 7.6.8 let C be an analytic Jordan curve. Then for $k + \alpha > 3/2$ we have

$$\left| f(z) - \sum_{\nu=0}^{n} b_\nu R_\nu(z) \right| \leqq \frac{M}{n^{k+\alpha-3/2}}, \quad z \text{ on } C_\rho .$$

SUGGESTION. Use the fact that $|R_n(z)| \leqq M n^{1/2} \rho^n$, z on C_ρ. [Carleman, 1922]

7.6.13. Let C consist of one or more Jordan curves each composed of a finite number of analytic arcs meeting in corners of exterior openings $\pi \leqq t\pi < 2\pi$. Let f(z) be defined on $\overline{C}$ and let $p_n(z)$ exist such that

$$\iint_C \Delta(z) \left| f(z) - p_n(z) \right|^p dS \leqq \frac{M}{n^\beta \rho^{np}}, \quad p > 0,$$

where $\Delta(z)$ is positive and continuous on C. Then we have

$$\left| f(z) - p_n(z) \right| \leqq \frac{M_1 r^n}{n^{(\beta-2t)/p} \rho^n}, \quad z \text{ on } C_r, \ r < \rho.$$

7.6.14. In Ex. 7.6.13 we also have

$$\left| f(z) - p_n(z) \right| \leqq \frac{M_2}{n^{(\beta-2t)/p-1}}, \quad z \text{ on } C_\rho ,$$

provided $(\beta-2t)/p > 1$.

7.6.15. In Ex. 7.6.13 let C_ρ consist of mutually exterior curves and suppose $(\beta-2t)/p-1 = k+\alpha > 0$. Then f(z) belongs to the class $L(k, \alpha)$ on C_ρ if $0 < \alpha < 1$, and to the class Log(k,1) on C_ρ if $\alpha = 1$.

§7.7. DISCUSSION. The results of this chapter depend to a large extent upon the methods and results of previous chapters; hence an extension of certain previous results, such as those of §§2.5, 3.2, and 5.1, would lead immediately to corresponding extensions here.

Just as in Problem α the surface integral theory of Problem β presents an opportunity worthy of serious consideration; certain ones of the exercises serve to emphasize this fact. We have merely to compare, or contrast, Exercises 7.6.1 and 7.6.7, or Exercises 7.6.1 and 7.6.15.

The generality of Theorem 7.2.1 is particularly gratifying; we require merely that C be a rectifiable Jordan curve or arc. It is to be observed that the extension of this theorem from one to several curves (Theorem 7.2.2) necessitates severe restrictions on the continuity properties of the curves. This is a striking instance of how an improvement of previous results (here Theorem 5.3.2) would lead directly to theorems on approximation for more general configurations.

The remarks in §6.8 concerning C_ρ with multiple points are pertinent in connection with the results of the present chapter.

CHAPTER VIII

SPECIAL CONFIGURATIONS

§8.1. APPROXIMATION ON $|z| = 1$ BY POLYNOMIALS IN z AND $1/z$. The special configurations which we consider in this chapter are (i) the circle, (ii) the segment $-1 \leq z \leq +1$, and (iii) the real axis in which case we study periodic functions and trigonometric approximation. As far as Problem α is concerned we have already studied (§§3.1, 3.2) approximation by polynomials in z and by trigonometric sums on these particular configurations. In the language of the real domain these results are classical and are due primarily to de la Vallée Poussin [1919] and Jackson [1930]. Some of the exercises at the end of the present chapter (§8.6) express these results in the language of the complex domain.

In our consideration of the unit circle we extend the results to polynomials in z and $1/z$. The purpose of this extension is not only to obtain new results on the theory of approximation in the complex domain but also to point out a connection with the theory of approximation in the real domain; as a matter of fact the theorems which we establish lie on the border lines of these two general fields of research.

Since on the unit circle a polynomial in z and $1/z$ is a trigonometric sum and conversely, it is quite natural to consider such polynomials. Walsh [1935, p.39] showed that a function $f(z)$ continuous on an arbitrary Jordan curve of the finite z-plane can be uniformly approximated by polynomials in z and $1/z$. This, of course, is a result on Problem α; theorems on degree of approximation (Problem α) by polynomials in z and $1/z$ are included in the exercises (§8.6).

Our main concern in the present chapter is Problem β, particularly in §§8.1, 8.2, and 8.3; the results of these three sections have the same general references. C. de la Vallée Poussin [1919] considered trigonometric approximation and functions analytic in a band containing the real axis; he studied particularly functions with poles and other isolated singularities on the boundary of the region of analyticity. Bernstein [1926] obtained results on degree of approximation on the line segment, in particular for functions with singularities on C_ρ (ellipse). The theorems as stated in these three sections (§8.1, 8.2, and 8.3) are due to Walsh and Sewell [1938, 1940a, 1941].

We proceed with the study of approximation on $|z| = 1$. For our present purposes we introduce a natural extension of the definition of a function of class $L(k, \alpha)$ or $L'(k, \alpha)$. Let $f(z)$ be analytic in the annulus γ_ρ: $\rho > |z| > 1/\rho < 1$; then we have the Laurent development $f(z) = \sum_{\nu=-\infty}^{\nu=\infty} c_\nu z^\nu$, $\rho > |z| > 1/\rho$. In the usual proof of the validity of this development it is established that we have for $\rho > |z| > 1/\rho$ the equation $f(z) \equiv F_1(z) + F_2(z)$, where

$$F_1(z) = \sum_{\nu=0}^{\infty} c_\nu z^\nu, \quad |z| < \rho \; ; \; F_2(z) = \sum_{\nu=-1}^{-\infty} c_\nu z^\nu, \quad |z| > 1/\rho.$$

If $F_1(z)$ and $F_2(1/z)$ belong to the class $L(k, \alpha)$ or $L'(k, \alpha)$ on $|z| = \rho$ we say that $f(z)$ belongs to the <u>class $L(k, \alpha)$ or $L'(k, \alpha)$ on γ_ρ</u>. With this definition we proceed to establish theorems analogous to those of Chapters VI and VII.

We prove first the direct Tchebycheff theorem.

THEOREM 8.1.1. Let $f(z)$ belong to the class $L(k, \alpha)$ or $L'(k, \alpha)$ on γ_ρ: $\rho > |z| > 1/\rho$ and let

$$f(z) = \sum_{\nu=-\infty}^{\infty} c_\nu z^\nu, \quad \rho > |z| > 1/\rho < 1.$$

Then with the notation $a_\nu = c_\nu + c_{-\nu}$, $b_\nu = i(c_\nu - c_{-\nu})$, we have on $|z| = 1$, with $z = e^{i\theta}$, the relation

$$(8.1.1) \quad \left| f(e^{i\theta}) - \left[\frac{a_0}{2} + \sum_{\nu=1}^{n} (a_\nu \cos \nu\theta + b_\nu \sin \nu\theta)\right]\right| \leqq \frac{M}{n^{k+\alpha}\rho^n},$$

for all θ.

We have merely to consider $F_1(z)$ and $F_2(z)$ separately and recall (Ex. 6.7.10) that we have

$$\left| F_1(z) - \sum_{\nu=0}^{n} c_\nu z^\nu \right| \leqq \frac{M_1}{n^{k+\alpha}\rho^n}, \quad |z| = 1,$$

$$\left| F_2(1/z) - \sum_{\nu=-1}^{-n} c_\nu z^\nu \right| \leqq \frac{M_2}{n^{k+\alpha}\rho^n}, \quad |1/z| = 1,$$

$$\text{or} \quad \left| F_2(z) - \sum_{\nu=-1}^{-n} c_\nu z^\nu \right| \leqq \frac{M_2}{n^{k+\alpha}\rho^n}, \quad |z| = 1;$$

thus

$$\left| f(e^{i\theta}) - \sum_{\nu=-n}^{n} c_\nu e^{\nu i\theta} \right| \leqq \frac{M_3}{n^{k+\alpha}\rho^n}.$$

But $e^{\nu i\theta} = \cos \nu\theta + i \sin \nu\theta$, $e^{-\nu i\theta} = \cos \nu\theta - i \sin \nu\theta$, hence we have

$$\left| f(e^{i\theta}) - c_0 - \sum_{\nu=1}^{n} [(c_\nu + c_{-\nu})\cos \nu\theta + i(c_\nu - c_{-\nu})\sin \nu\theta] \right| \leqq \frac{M}{n^{k+\alpha}\rho^n}.$$

Inequality (8.1.1) now follows by the definition of the a_ν and b_ν. The proof of the theorem is complete.

The direct theorem for approximation in the sense of

least p-th powers is an immediate consequence of the above result.

THEOREM 8.1.2. Under the hypothesis of Theorem 8.1.1 we have

$$\int_{-\pi}^{\pi} \Delta(\theta)\left|f(e^{i\theta}) - \left[\frac{a_0}{2} + \sum_{\nu=1}^{n} (a_\nu \cos \nu\theta + b_\nu \sin \nu\theta)\right]\right|^p d\theta$$

$$\leq \frac{M}{n^{k+\alpha}\rho^n}, \quad p > 0,$$

where $\Delta(\theta)$ is a positive and continuous function of θ, periodic of period 2π.

In the converse direction we have

THEOREM 8.1.3. Let $f(\theta)$ be periodic with period 2π, and suppose the numbers $a_{n\nu}$ and $b_{n\nu}$ (not necessarily real) are given so that

$$s_n(\theta) = \frac{a_{n0}}{2} + \sum_{\nu=1}^{n} (a_{n\nu} \cos \nu\theta + b_{n\nu} \sin \nu\theta),$$

with the relation, for $n = 1, 2, \ldots$, and for all θ,

$$(8.1.2) \quad |f(\theta) - s_n(\theta)| \leq \frac{M}{n^{k+\alpha+1}\rho^n}, \quad \rho > 1.$$

Then the function

$$(8.1.3) \quad F(z) = \lim_{n\to\infty} \left[c_{n0} + \sum_{\nu=0}^{n} (c_{n,-\nu} z^{-\nu} + c_{n\nu} z^\nu)\right],$$

$$2c_{n\nu} = a_{n\nu} - ib_{n\nu}, \quad 2c_{n,-\nu} = a_{n\nu} + ib_{n\nu},$$

coincides with $f(\theta)$ on the circle $|z| = 1$, with $z = \cos\theta + i \sin\theta$ and $F(z)$ belongs to the class $L(k, \alpha)$ on γ_ρ if $k + \alpha + 1$ is not a positive integer, and to the class $L'(k, 1)$ on γ_ρ if $k + \alpha + 1$ is a positive integer.

Inequality (8.1.2) written for successive indices implies

$$|s_{n+1}(\theta) - s_n(\theta) \leq \frac{M_1}{n^{k+\alpha+1}\ \rho^n}$$

for all θ, an inequality which we write in the form

$$(8.1.4)\quad \left|p_{n+1}(z,\ 1/z) - p_n(z,\ 1/z)\right| \leq \frac{M_1}{n^{k+\alpha+1}\ \rho^n}\ , |z| = 1,$$

where $p_n(z, 1/z)$ is the expression in square brackets in (8.1.3), a polynomial in z and $1/z$ of degree n, equal to $s_n(\theta)$ on $|z| = 1$. In order to proceed with the method of proof of Theorem 6.3.1 we need the following lemma due to Walsh [1935, p. 259]:

LEMMA 8.1.4. Let Γ be an arbitrary Jordan curve of the z-plane in whose interior the origin lies. If $p_n(z, 1/z)$ is a function of the form

$$p_n(z, 1/z) = a_{-n}\ z^{-n} + a_{-n+1}\ z^{-n+1} + \ \dots\ + a_0 + a_1 z + \dots + a_n z^n,$$

and if we have, for z on Γ

$$|p_n(z, 1/z)| \leq M,$$

then we have also

$$(8.1.5)\quad \left|p_n(z, 1/z)\right| \leq M\ \rho^n, \quad \rho > 1,$$

for z on $\bar{\Gamma}_\rho$. Here $\bar{\Gamma}_\rho$ denotes the closed region between and bounded by the two curves $|\phi(z)| = \rho$ and $|\Phi(z)| = \rho$, where $w = \phi(z)$ maps the exterior of Γ onto $|w| > 1$ so that $z = \infty$ corresponds to $w = \infty$, and $w = \Phi(z)$ maps the interior of Γ onto $|w| > 1$ so that $z = 0$ corresponds to $w = \infty$.

We use a method similar to that employed in the proof of Theorem 2.1.3. The function

$$p_n(z, 1/z)\ [\phi(z)]^{-n}$$

is analytic exterior to Γ, continuous in the corresponding closed region. Its modulus on Γ is not greater than M, so its modulus on $|\phi(z)| = \rho$ is also not greater than M. That is to say, inequality (8.1.5) is valid on $|\phi(z)| = \rho$. Similarly we derive (8.1.5) for z on $|\Phi(z)| = \rho$. The function $p_n(z,1/z)$ is analytic in the closed region $\overline{\Gamma}_\rho$; inequality (8.1.5) is valid on the boundary, hence valid in the closed region.

An application of the lemma to inequality (8.1.4) yields

$$|p_{n+1}(z,1/z)-p_n(z,1/z)| \leq \frac{M_1 r^n}{n^{k+\alpha+1}\rho^n}, \quad 1 < r < \rho,$$

$$r \geq |z| \geq 1/r.$$

If $k + \alpha < 0$ we proceed directly as in the proof of Theorem 6.3.1. We obtain then $|F(z)| \leq M(r-1)^{k+\alpha}$, for $|z| = r$ and $|z| = 1/r$. If we write as above $F(z) = F_1(z) + F_2(z)$, we see that $|F_1(z)| \leq M(r-1)^{k+\alpha}$, $|z| = r$, since $F_2(z)$ is analytic and uniformly bounded in $|z| \geq 1$; hence $F_1(z)$ belongs to the class $L(k, \alpha)$ on $|z| = \rho$. Likewise $F_2(1/z)$ belongs to the class $L(k, \alpha)$ on $|z| = \rho$. Thus by definition $F(z)$ belongs to the class $L(k, \alpha)$ on γ_ρ.

If $k+\alpha \geq 0$ we need a further lemma due to the author [1938b]:

LEMMA 8.1.5. Let Γ be a Jordan curve of Type t, $1 \leq t \leq 2$, in whose interior the origin lies. Then $|p_n(z,1/z)| \leq M$, z on Γ, implies $|p_n'(z,1/z)| \leq K(\Gamma)Mn^t$, z on Γ, where $K(\Gamma)$ is a constant depending only on t and Γ.

The proof of this lemma is easy; the method of proof of Theorem 2.1.4 suffices. We have merely to note that Theorem 2.1.3 is valid for interior as well as exterior level curves; that is to say, (in the notation of Lemma 8.1.4) for the curves $|\Phi(z)| = \rho$ as well as for the curves $|\phi(z)| = \rho$. The proof is completed by applying Lemma 8.1.4 instead of Theorem 2.1.3. The details are left to the reader. Of course in the present situation $t = 1$.

For the case of Γ the unit circle this lemma is a weak form of Bernstein's Lemma; it is a weak form in the sense that the multiplicative constant $K(\Gamma)$ is not the best possible. However this weaker form is sufficient for our present purposes.

With the aid of this lemma it is easy to establish the inequality

$$|F'(z) - p_n'(z,1/z)| \leqq \frac{M_1}{n^{k+\alpha+1-1}\rho^n}, \quad |z| = 1,$$

by the method of Theorem 6.2.8. Thus for suitable p we know from the discussion for $k+\alpha < 0$ that $F^{(p)}(z)$ belongs to the class $L(k-p, \alpha)$, $k-p<0$, on γ_ρ. Then by integrating p times and applying Theorem 5.2.8 or Definition 5.2.3 we have the conclusion of the theorem. The proof is complete.

Our indirect integral theorem is

THEOREM 8.1.6. Let $f(\theta)$ be periodic with period 2π, and suppose the numbers $a_{n\nu}$ and $b_{n\nu}$ (not necessarily real) are given so that

$$s_n(\theta) = \frac{a_{no}}{2} + \sum_{\nu=1}^{n} (a_{n\nu} \cos \nu\theta + b_{n\nu} \sin \nu\theta),$$

with the relation, for $n = 1, 2, \ldots,$

$$(8.1.6) \quad \int_{-\pi}^{\pi} \Delta(\theta)|f(\theta) - s_n(\theta)|^p d\theta \leqq \frac{M}{n^{(k+\alpha+1)p}\rho^{np}},$$

$$p > 0,$$

where $\Delta(\theta)$ is a positive and continuous function of θ, periodic of period 2π. Then the function (8.1.3) coincides with $f(\theta)$ almost everywhere on $|z| = 1$, with $z = \cos\theta + i\sin\theta$, and $F(z)$ belongs to the class $L(k, \alpha)$ on γ_ρ if $k + \alpha + 1$ is not a positive integer, and to the class $L'(k, \alpha)$ on γ_ρ if $k + \alpha + 1$ is a positive integer.

We use the method of proof of Theorem 7.2.1. As there we have by virtue of (8.1.6)

$$\int_{-\pi}^{\pi} |s_{n+1}(\theta) - s_n(\theta)|^p d\theta \leq \frac{M_1}{n^{(k+\alpha+1)p}\rho^{np}},$$

or

$$(8.1.7) \qquad \int_{|z|=1} |p_{n+1}(z,1/z)-p_n(z,1/z)|^p |dz| \leq \frac{M_1}{n^{(k+\alpha+1)p}\rho^{np}}.$$

Our method now requires the following lemma:

LEMMA 8.1.7. Let Γ be a rectifiable Jordan curve in whose interior the origin lies. Let

$$\int_\Gamma |p_n(z,1/z)|^p |dz| \leq L^p, \quad p > 0.$$

Then (in the notation of Lemma 8.1.4) we have

$$(8.1.8) \quad |p_n(z,1/z)| \leq KL\rho^n, \quad z \text{ on } \bar{\Gamma}_\rho,$$

where K is a constant depending on Γ, ρ, and p, but independent of $p_n(z,1/z)$, n, and z.

By the method of proof of Theorem 5.3.1 we have inequality (8.1.8) for z on $|\phi(z)| = \rho$. Then by considering the interior map (see proof of Lemma 8.1.4) and using precisely the same method we have inequality (8.1.8) for z on $|\Phi(z)| = \rho$. Inequality (8.1.8) for z on $\bar{\Gamma}_\rho$ then follows by virtue of the analyticity of $p_n(z,1/z)$ on $\bar{\Gamma}_\rho$ and the principle of the maximum.

Inequality (8.1.7) now yields by virtue of Lemma 8.1.7

$$|p_{n+1}(z)-p_n(z)| \leq \frac{M_2 r^n}{n^{k+\alpha+1}\rho^n}, \quad 1 < r < \rho , \quad r \geq |z| \geq 1/r.$$

The theorem follows by the argument subsequent to the corresponding inequality in the proof of Theorem 8.1.3; the equality of $f(\theta)$ and $F(z)$ almost everywhere on $|z| = 1$ follows by the usual reasoning (§4.2).

Examples similar to those used in connection with approximation by polynomials in z serve to show that these results are the best possible in the usual sense.

§8.2. APPROXIMATION ON $-1 \leq z \leq 1$. By means of a suitable conformal transformation we are able to apply the results of the preceding section to approximation on the segment $-1 \leq z \leq 1$.

Analogous to Theorem 8.1.1 we have

THEOREM 8.2.1. Let Γ_ρ denote the ellipse whose foci are -1 and $+1$ and whose sum of semi-axes is ρ, and let $f(z)$ belong to the class $L(k, \alpha)$ or $L'(k, \alpha)$ on Γ_ρ. Then $P_n(z)$ exists such that

$$|f(z)-P_n(z)| \leq \frac{M}{n^{k+\alpha}\rho^n}, \quad -1 \leq z \leq 1.$$

We map the z-plane onto the w-plane by the transformation (see §3.4) $z=(w+w^{-1})/2$. Under this transformation the image in the w-plane of the segment $-1 \leq z \leq 1$ counted twice is the unit circle $|w| = 1$, the image in the w-plane of Γ_ρ in the z-plane counted twice consists of the two circles $|w| = \rho$ and $|w| = 1/\rho$, and the image in the w-plane of the interior of Γ_ρ in the z-plane counted twice is the annular region $\rho > |w| > 1/\rho$. The transformation $z = (w+w^{-1})/2$ is analytic on $|w| = \rho$ and

$|w| = 1/\rho$.

This is the transformation $z = \cos\theta$ which we used in the proof of Theorem 3.2.3, for on $|w| = 1$ we have

$$z = \frac{1}{2}(w + \frac{1}{w}) = \frac{1}{2}(\cos\theta + i\sin\theta + \frac{1}{\cos\theta + i\sin\theta}) = \cos\theta.$$

This transformation is well behaved on $|w| = 1$ except at the points $w = 1$ and $w = -1$ which correspond respectively to the points $z = 1$ and $z = -1$; it is to be observed that the segment (counted twice) $-1 \leqq z \leqq 1$ is analytic for $z^2 \leqq a < 1$, but has corners at $+1$ and -1 whose exterior openings are 2π. A careful study of this transformation is enlightening as far as Theorem 3.2.3 and Theorem 3.2.1, applied to this particular case, are concerned.

Under this transformation the function $f(z)$ corresponds to a function of w which is analytic in the neighborhood of $|w| = 1$ except perhaps on the circumference, continuous in the two-dimensional sense at every point of $|w| = 1$, and hence analytic on $|w| = 1$ and throughout the annulus $\rho > |w| > 1/\rho$. The function $f[(w+w^{-1})/2]$ is symmetric in w and $1/w$, so that the corresponding Laurent polynomials used in the proof of Theorem 8.1.1 are in this case symmetric in w and $1/w$ and hence are polynomials in z; Theorem 8.2.1 now follows from Theorem 8.1.1.

The analogue of Theorem 8.1.2 is an immediate consequence of Theorem 8.2.1.

THEOREM 8.2.2. Under the hypothesis of Theorem 8.2.1, let $\Delta(z)$ be positive and continuous on $-1 \leqq z \leqq 1$; then there exists $P_n(z)$ such that we have

$$\int_{-1}^{1} \Delta(z)|f(z)-P_n(z)|^p dz \leqq \frac{M}{n^{(k+\alpha)p}\rho^{np}}, \quad p > 0.$$

The indirect Tchebycheff theorem is referred back to

Theorem 8.1.3 by the same transformation.

THEOREM 8.2.3. Let $f(z)$ be defined on the segment $-1 \leqq z \leqq 1$, and let $P_n(z)$ exist such that

$$|f(z)-P_n(z)| \leqq \frac{M}{n^{k+\alpha+1}\rho^n}, \quad -1 \leqq z \leqq 1.$$

Then the function $f(z)$, if suitably defined, belongs to the class $L(k, \alpha)$ on Γ_ρ if $k+\alpha+1$ is not a positive integer, and to the class $L'(k, \alpha)$ on Γ_ρ if $k+\alpha+1$ is a positive integer, where Γ_ρ is the ellipse whose foci are -1 and $+1$ and whose sum of semi-axes is ρ.

Under the transformation $z = (w+w^{-1})/2$ the function $f(z)$ corresponds to a function $F(w)$ defined on $|w| = 1$; the polynomial $P_n(z)$ corresponds to a polynomial in w and $1/w$ of degree n. Thus the hypothesis of Theorem 8.1.3 is fulfilled and hence the function $F(w)$ belongs to the class $L(k, \alpha)$ or $L'(k, \alpha)$, as the case may be, on γ_ρ. Thus the function $f(z)$, the transform of $F(w)$, belongs to the class $L(k, \alpha)$ or $L'(k, \alpha)$ on the ellipse Γ_ρ by virtue of the analyticity of the transformation on $|w| = \rho$ and $|w| = 1/\rho$ and the argument presented above.

The indirect theorem analogous to Theorem 8.1.6 is closely related to Theorem 7.2.1 and will not be stated explicitly here. It is interesting to note that under the above transformation inequality (7.2.1) becomes

$$\int_{|w|=1} \Delta(z)|f(z)-P_n(z)|^p |1-z^2|^{1/2}|dw| \leqq \frac{M}{n^{(k+\alpha+1)p}\rho^{np}},$$

and hence to satisfy the hypothesis of Theorem 8.1.6 it is necessary to assume $\Delta(z)|1-z^2|^{1/2}$, rather than as in Theorem 7.2.1 $\Delta(z)$ itself, positive and continuous on $-1 \leqq z \leqq 1$.

Even though the segment $-1 \leq z \leq 1$ is not an analytic Jordan curve we do have [Walsh and Sewell, 1941] a result similar to Theorems 6.2.3 and 6.2.4. The roots of the polynomial $T_n(z) = 2^{-n+1} \cos(n \cos^{-1} z)$ are equally distributed on the segment C: $-1 \leq z \leq +1$, as is easily seen by considering the transformation $z = (w+w^{-1})/2$; since $z = \cos\theta$ on $|w| = 1$ we have $T_n(z) = 2^{-n+1} \cos n\theta$, whose roots are obviously equally distributed on $|w| = 1$. We have (§6.7) $|T_n(z)/T_n(t)| \leq M/\rho^n$, z on C, t on C_ρ. Let $f(z)$ belong to the class $L(k, \alpha)$ on C_ρ, whence we have by formula (5.1.2)

$$f(z) - p_{n-1}(z) = \frac{1}{2\pi i} \int_{C_r} \frac{T_n(z)}{T_n(t)} \cdot \frac{f(t)-P_{n-1}(t)}{t - z}\, dt,$$

$$-1 \leq z \leq 1, \quad 1 < r < \rho .$$

Integration yields

$$\int_0^z [f(z)-p_{n-1}(z)]dz = \frac{1}{2\pi i} \int_{C_r} \frac{[f(t)-P_{n-1}(t)}{T_n(t)} \left[\int_0^z \frac{T_n(z)}{t-z} dz \right] dt,$$

whence we need to consider merely

$$\int_0^z \frac{\cos (n \cos^{-1} z)\, dz}{t - z} = \frac{1}{t-z} \int \cos (n \cos^{-1} z)\, dz \Bigg]_0^z$$

$$- \int_0^z \left[\int \cos (n \cos^{-1} z)\, dz \right] \frac{dz}{(t-z)^2} .$$

But

$$\int \cos (n \cos^{-1} z) dz = \frac{1}{2n} \left[\frac{n}{n+1} \cos[(n+1)\cos^{-1} z] \right.$$

$$\left. + \frac{n}{n+1} \cos[(n+1 \cos^{-1} z] \right] ;$$

thus for all t on C_r and z on C we have

$$\left| \int_0^z \frac{\cos (n \cos^{-1} z)}{t-z} dz \right| \leq \frac{M}{n} .$$

Consequently we obtain the same inequalities for integration of sequences in the case of a segment C as for C an analytic Jordan curve.

§8.3. TRIGONOMETRIC APPROXIMATION. Another conformal transformation enables us to obtain further results by virtue of the theorems of §8.1. We use now the transformation $w = e^{iz}$, $z = x + iy$, which carries the line $y = 0$ into the unit circle $|w| = 1$ and the band $|y| < \log \rho > 0$ into the annulus $\rho > |w| > 1/\rho$. We say that a function $f(z)$ periodic of period 2π belongs to the class $L(k, \alpha)$ or $L'(k, \alpha)$ in the band $|y| < \log \rho$ provided its transform $F(w)$ belongs to the class $L(k, \alpha)$ or $L'(k, \alpha)$, respectively, on the annulus $\rho > |w| > 1/\rho$. It follows that if $k \geq 0$ the function $f(z)$ is analytic in the band $|y| < \log \rho$, continuous in the corresponding closed region, and on the lines $y = \pm \log \rho$ we have

$$\left|f^{(k)}(z_1) - f^{(k)}(z_2)\right| \leq L|z_1-z_2|^{\alpha} \left|\log |z_1-z_2|\right|^{\beta},$$

where $\beta = 0$ for all α if $f(z)$ belongs to the class $L(k, \alpha)$, and $\beta = \alpha = 1$ if $f(z)$ belongs to the class $L'(k, \alpha)$.

We are now ready to prove the analogue of Theorem 8.1.1.

THEOREM 8.3.1. Let $f(z)$ be periodic of period 2π, and let $f(z)$ belong to the class $L(k, \alpha)$ or $L'(k, \alpha)$ in the band $|y| < \log \rho$, where $z = x + iy$. Then trigonometric sums (see §3.1) $t_n(z)$ of respective orders n exist such that for all real z we have

$$(8.3.1) \quad |f(z) - t_n(z)| \leq \frac{M}{n^{k+\alpha} \rho^n}.$$

Under the transformation $w = e^{iz}$ the function $F(w)$ belongs to the class $L(k, \alpha)$ or $L'(k, \alpha)$ in the annulus

$\rho > |w| > 1/\rho$, hence by Theorem 8.1.1 there exists $p_n(w, 1/w)$ such that

$$|F(w) - p_n(w, 1/w)| \leqq \frac{M}{n^{k+\alpha} \rho^n}, \quad |w| = 1.$$

But if $w = e^{iz}$, then $w^{\nu} = \cos \nu z + i \sin \nu z$, $w^{-\nu} = \cos \nu z - i \sin \nu z$, hence $p_n(w, 1/w)$ is a trigonometric sum $t_n(z)$ of order n; thus we have inequality (8.3.1) and the proof is complete.

An immediate consequence of this theorem is

THEOREM 8.3.2. Under the hypothesis of Theorem 8.3.1 let $\Delta(z)$ be a positive continuous function of period 2π. Then we have

$$\int_{-\pi}^{\pi} \Delta(z)|f(z)-t_n(z)|^p dz \leqq \frac{M}{n^{(k+\alpha)p} \rho^{np}}, \quad p > 0,$$

Our indirect Tchebycheff theorem is

THEOREM 8.3.3. Let f(z) be periodic of period 2π, and let trigonometric sums $t_n(z)$ of respective orders n exist such that for all real $z = x + iy$ we have

$$(8.3.2) \quad \left|f(z) - t_n(z)\right| \leqq \frac{M}{n^{k+\alpha+1} \rho^n}.$$

Then the function f(z) can be analytically extended so that it belongs to the class $L(k, \alpha)$ in the band $|y| < \log \rho$ if $k + \alpha + 1$ is not a positive integer, and to the class $L'(k, \alpha)$ if $k + \alpha + 1$ is a positive integer.

The proof of this theorem is readily supplied by the reader by use of the same transformation and the method of proof of Theorem 8.2.3.

The proof of the following theorem is also left to the reader:

THEOREM 8.3.4. Let $f(z)$ be periodic of period 2π, and let trigonometric sums $t_n(z)$ of respective orders n exist such that we have

$$\int_{-\pi}^{\pi} \Delta(z)\left|f(z) - t_n(z)\right|^p dz \leq \frac{M}{n^{(k+\alpha)p}\,\rho^{np}}, \quad p > 0,$$

where $\Delta(z)$ is a positive continuous function of period 2π. Then if we define $F(z) = \lim_{n\to\infty} t_n(z)$, where $t_n(z)$ is still expressed as a trigonometric sum, the two functions $f(z)$ and $F(z)$ are equal almost everywhere on the axis of reals. Moreover if $z = x + iy$, the function $F(z)$ belongs to the class $L(k, \alpha)$ in the band $|y| < \log \rho$ if $k + \alpha + 1$ is not a positive integer, and to the class $L'(k, \alpha)$ in the band $|y| < \log \rho$ if $k + \alpha + 1$ is a positive integer.

Of course, in this theorem as well as in the other integral theorems of this and the two preceding sections the weight function may be generalized according to the discussion of §7.5.

§8.4. DIRECT METHODS ON PROBLEM α AND PROBLEM β. As the title indicates, the emphasis in the present section is on the methods rather than the results. In fact we have already established results, both on Problem α and Problem β, which in many respects are as favorable as can be obtained; however, our methods have been in general relatively high-powered. For instance our proof of Theorem 3.2.1 of Problem α even for the unit circle is based on Theorem 3.1.1 with its highly complicated proof; in fact the real variable method does not even yield an elementary proof that $\left|f(z) - \sum_{\nu=0}^{n} a_\nu z^\nu\right| \leq M(\log n/n^\alpha)$, $|z| \leq 1$, where $f(z)$ belongs to the class $L(0, \alpha)$ on $|z| = 1$. In Problem β

it is true that our proof of Theorem 6.1.1 is elementary provided $k + \alpha < 0$; however for $k + \alpha \geq 0$ we must use either Theorem 3.2.1 or the relatively involved results on differentiation and integration. Nevertheless, some results only slightly less favorable than those previously obtained can be established by thoroughly immediate and elementary methods, with a minimum of machinery, as we now proceed to indicate.

For Problem α we have

THEOREM 8.4.1. Let $f(z)$ belong to the class $L(0, \alpha)$ on $|z| = 1$. Then $p_n(z)$ exists such that

$$(8.4.1) \quad |f(z)-p_n(z)| \leq \frac{M(\log n)^\alpha}{n^\alpha}, \quad |z| \leq 1;$$

indeed, $p_n(z)$ may be defined as $s_n(r_n z)$, where $s_n(z)$ is the sum of the first $n+1$ terms of the Taylor development of $f(z)$ and $r_n = 1 - (2\log n)/n$.

We set

$$f(z) = \sum_{\nu=0}^{\infty} a_\nu z^\nu, \quad s_n(z) = \sum_{\nu=0}^{n} a_\nu z^\nu,$$

and we have

$$(8.4.2) \quad |f(re^{i\theta}) - s_n(re^{i\theta})| = \left| \frac{1}{2\pi i} \int_{|t|=1} \frac{(re^{i\theta})^{n+1}}{t^{n+1}} \frac{f(t)}{t-z} dt \right|$$

$$\leq \frac{M_1 r^{n+2}}{1-r}, \quad r < 1, \text{ for all } \theta,$$

where M_1 is a constant depending only on $f(z)$. Since $f(z)$ belongs to the class $L(0, \alpha)$ we have

$$(8.4.3) \quad |f(e^{i\theta}) - f(re^{i\theta})| \leq L(1-r)^\alpha, \quad 0 < r < 1,$$

for all θ; hence for θ arbitrary but fixed we have by addition of (8.4.2) and (8.4.3)

$$\left|f(e^{i\theta}) - s_n(re^{i\theta})\right| \leq \frac{M_1 r^{n+2}}{1-r} + L(1-r)^{\alpha} .$$

Corresponding to each n we choose

$$r = r_n = 1 - \frac{2 \log n}{n};$$

but

$$(1-\frac{2 \log n}{n})^n = \left[(1- \frac{2 \log n}{n})^{n/2 \log n}\right]^{2 \log n}$$

is asymptotic to n^{-2} in the sense that $(1-(2 \log n)/n)^n/n^2$ is bounded. Thus we obtain inequality (8.4.1) for $|z| = 1$. The validity of (8.4.1) for $|z| \leq 1$ follows by the principle of the maximum.

It is to be observed that we use the Lipschitz inequality simply on the radius rather than in the closed circle.

It is easy to apply Theorem 8.4.1 to Problem β:

THEOREM 8.4.2. Let f(z) satisfy the hypothesis of Theorem 8.4.1. Then $p_n(z)$ exists such that

$$(8.4.4) \quad |f(z)-p_n(z)| \leq \frac{M(\log n)^{\alpha}}{n^{\alpha} \quad \rho^n}, \quad |z| \leq 1/\rho < 1;$$

indeed $p_n(z)$ is the sum of the first n+1 terms of the Taylor development of f(z).

We write

$$f(z)-p_n(z) = \frac{1}{2\pi i} \int_{|t|=1} \frac{z^{n+1} f(t)}{t^{n+1}(t-z)} dt$$

$$= \frac{1}{2\pi i} \int_{|t|=1} \frac{z^{n+1} f(rt)}{t^{n+1}(t-z)} dt$$

$$+ \frac{1}{2\pi i} \int_{|t|=1} \frac{z^{n+1}[f(t)-f(rt)]}{t^{n+1}\ (t-z)} dt, 0 < r < 1.$$

For each n we choose $r = r_n = 1 - \log n/n$, whence by the method used in Theorem 8.4.1 we have the inequality of the theorem.

The method of Theorem 8.4.2 extends in an elementary way to yield results on approximation to integrals and derivatives of f(z), hence to include all values of k; compare §6.2.

By using the Lipschitz inequality on the circumference rather than along a radius we obtain a stronger result; in fact we now present an elementary proof of Theorem 6.1.1 for a function f(z) belonging to the class $L(0, \alpha)$ on $|z| = 1$. Our problem is essentially one in Fourier coefficients and we employ a standard method of that theory. If $f(z) = \sum_{\nu=0}^{\infty} a_\nu z^\nu$ we may write [see, e.g., Zygmund, 1935]

$$a_n = \frac{1}{2\pi i} \int_{|z|=1} \frac{f(z)}{z^{n+1}}\, dz = \frac{1}{2\pi} \int_0^{2\pi} \frac{f(e^{i\theta})}{e^{ni\theta}}\, d\theta$$

$$= -\frac{1}{2\pi} \int_0^{2\pi} \frac{f(e^{i(\theta+\pi/n)})}{e^{ni\theta}}\, d\theta$$

$$= \frac{1}{2\pi} \int_0^{2\pi} [f(e^{i\theta}) - f(e^{i(\theta+\pi/n)})] \frac{d\theta}{e^{ni\theta}},$$

whence we see that $|a_n| \leq M/n^\alpha$. Consequently on the circle $|z| = 1/\rho < 1$ we have

$$\left| f(z) - \sum_{\nu=0}^{n} a_\nu z^\nu \right| \leq \sum_{\nu=n+1}^{\infty} |a_\nu z^\nu| \leq M \sum_{\nu=n+1}^{\infty} \frac{1}{\nu^\alpha \rho^\nu}$$

$$\leq \frac{M_1}{n^\alpha \rho^n};$$

thus we have proved Theorem 6.1.1 for $|z| = 1$ and $k = 0$. If f(z) is of class $L(k, \alpha)$, $k > 0$ we have by k-fold differentiation and use of the preceding relation $|a_n| \leq M/n^{k+\alpha}$, which extends our proof to include all $k \geq 0$. Since the proof of Theorem 6.1.1 for $k < 0$ is elementary we thus have an elementary proof of Theorem 6.1.1 for C a circle and for all k. We may extend the result from $k = 0$ to $k > 0$ by integration.

The material of the present section is due to Walsh and Sewell [1941].

§8.5. EXERCISES. 8.5.1. Let f(z) be defined on $|z| = 1$; let $f^{(k)}(z)$ exist there and satisfy a Lipschitz condition of order α. Then $p_n(z, 1/z)$ (in the notation of §8.1) exists such that

$$(8.5.1) \quad \left|f(z) - p_n(z, 1/z)\right| \leq \frac{M}{n^{k+\alpha}}, \quad |z| = 1.$$

This is essentially Theorem 3.1.1. [Jackson, 1930]

8.5.2. Let f(z) be defined for all real z, and let f(z) be periodic of period 2π; let $f^{(k)}(z)$ exist and satisfy a Lipschitz condition of order α. Then $t_n(z)$ (notation of §8.3) exists such that for all real z

$$(8.5.2) \quad \left|f(z) - t_n(z)\right| \leq \frac{M}{n^{k+\alpha}}.$$

This is essentially Theorem 3.1.1. [Jackson, 1930]

8.5.3. Let f(z) be defined on $|z| = 1$ and let $p_n(z, 1/z)$ exist such that (8.5.1) is valid on $|z| = 1$ for $k \geq 0$. Then f(z) belongs to the class $L(k, \alpha)$ on $|z| = 1$ if $\alpha < 1$, and to the class $Log(k, 1)$ on $|z| = 1$ if $\alpha = 1$. [de la Vallée Poussin, 1919]

8.5.4. Let f(z) be defined on the real axis and let $t_n(z)$ exist such that (8.5.2) is valid for all real z with $k \geq 0$. Then f(z) is continuous and periodic of period 2π; furthermore $f^{(k)}(z)$ exists and satisfies the condition

$$(8.5.3) \quad \left|f^{(k)}(z_1) - f^{(k)}(z_2)\right| \leq L|z_1 - z_2|^{\alpha} \left|\log|z_1 - z_2|\right|^{\beta},$$

for all real z, where $\beta = 1$ if $\alpha = 1$ and $\beta = 0$ if $\alpha < 1$, and L is a constant independent of z_1 and z_2.

8.5.5. Under the conditions of Ex. 8.5.1 we have

$$\int_{|z|=1} \Delta(z)|f(z)-p_n(z,1/z)|^p|dz| \leq \frac{M}{n^{(k+\alpha)p}}, \quad p > 0,$$

where $\Delta(z)$ is (here and in what follows) positive and continuous on $|z| = 1$.

8.5.6. Under the hypothesis of Ex. 8.5.2 we have

$$\int_{-\pi}^{\pi} \Delta(z)|f(z)-t_n(z)|^p \, dz \leq \frac{M}{n^{(k+\alpha)p}}, \quad p > 0.$$

8.5.7. Let $f(z)$ be defined on $|z| = 1$ and let $p_n(z,1/z)$ exist such that

$$\int_{|z|=1} \Delta(z)|f(z)-p_n(z,1/z)|^p|dz| \leq \frac{M}{n^{(k+\alpha)p+1}}, \quad p > 0.$$

Then $f(z) = f_1(z)$ almost everywhere on $|z| = 1$, where

$$|f_1(z) - p_n(z,1/z)| \leq \frac{M_1}{n^{k+\alpha}}, \quad |z| = 1;$$

furthermore $f_1^{(k)}(z)$ exists on $|z| = 1$ and satisfies condition (8.5.3) there.

8.5.8. Let $f(z)$ be defined on the real axis and let $t_n(z)$ exist such that

$$\int_{-\pi}^{\pi} \Delta(z)|f(z) - t_n(z)|^p dz \leq \frac{M}{n^{(k+\alpha)p+1}}, \quad p > 0.$$

Then $f(z) = f_1(z)$ almost everywhere on the real axis, where for all real z

$$|f_1(z) - t_n(z)| \leq \frac{M}{n^{k+\alpha}};$$

furthermore $f_1^{(k)}(z)$ exists and satisfies condition (8.5.3) for all real z.

8.5.9. Let $f(z)$ be defined on the segment E: $-1 \leq z \leq 1$ and let $P_n(z)$ exist such that

$$\int_{-1}^{1} \Delta(z)|f(z)-P_n(z)|^p dz \leq \frac{M}{n^{(k+\alpha)p+1}}, \quad p > 0.$$

Then $f(z) = f_1(z)$ almost everywhere on E, where $f_1(z) = \lim_{n \to \infty} P_n(z)$, z on E; furthermore in any interior interval (a,b), $-1 < a \leq z \leq b < 1$, the derivative $f_1^{(k)}(z)$ exists and satisfies condition (8.5.3).

8.5.10. In Theorem 8.1.1 we have for $k > 0$

$$\left| f(z) - \sum_{\nu=-n}^{\nu=n} c_\nu z^\nu \right| \leq \frac{M_1 \log n}{n^{k+\alpha}}, \quad \rho \geq |z| \geq 1/\rho \ .$$

[Walsh and Sewell, 1938]

8.5.11. In Theorem 8.3.1 we have for $k > 0$

$$\left| f(z) - t_n(z) \right| \leq \frac{M_1 \log n}{n^{k+\alpha}}, \quad |y| \leq \log \rho \ .$$

[Walsh and Sewell, 1938]

8.5.12. Under the hypothesis of Theorem 8.1.3 or Theorem 8.1.6 we have

$$\left| F(z) - p_n(z, 1/z) \right| \leq \frac{M_1 r^n}{n^{k+\alpha+1} \rho^n}, \quad r \geq |z| \geq 1/r, \ r < \rho \ .$$

Study the degree of convergence of polynomials in z and 1/z of best approximation here and below for both Problem β and Problem γ.

8.5.13. Under the hypothesis of Theorems 8.3.3 or 8.3.4 we have for z in the region $|y| < \log r < \log \rho$

$$\left| F(z) - t_n(z) \right| \leq \frac{M_1 r^n}{n^{k+\alpha+1} \rho^n} \ .$$

8.5.14. Under the hypothesis of Theorem 8.2.3 we have for z on and within the ellipse whose foci are -1 and $+1$ and whose sum of semi-axes is $r < \rho$,

$$\left| f(z) - P_n(z) \right| \leq \frac{M r^n}{n^{k+\alpha+1} \rho^n} \ .$$

8.5.15. Let E be the segment $-1 \leq z \leq 1$, so that C_ρ is the ellipse with foci -1 and $+1$ and with sum of semi-axes $\rho > 1$. Let $f(z)$ belong to the class $L(k, \alpha)$ on

C_ρ. Let $p_n(z)$ be the polynomial which interpolates to $f(z)$ in the n+1 roots of the Tchebycheff polynomial $T_{n+1}(z) = \cos[(n+1)\cos^{-1} z]/2^n$ belonging to E. Then we have

$$|f(z) - p_n(z)| \leq \frac{M}{n^{k+\alpha}\rho^n}, \quad -1 \leq z \leq 1.$$

[Walsh and Sewell, 1940]

SUGGESTION. The zeros of $T_n(z)$ lie on E, and

$$\lim_{n\to\infty} \frac{2^n T_n(z)}{[\phi(z)]^n} = 1,$$

uniformly on any closed set exterior to E.

8.5.16. Let E and $f(z)$ satisfy the hypothesis of Ex. 8.5.15, and let $p_n(z)$ be the polynomial which interpolates to $f(z)$ in the n+1 roots of the Legendre polynomial $L_{n+1}(z)$. Then we have

$$|f(z) - p_n(z)| \leq \frac{M}{n^{k+\alpha-1/2}\rho^n}, \quad -1 \leq z \leq 1.$$

[Walsh and Sewell; 1940; compare Shohat, 1933, 1934]

SUGGESTION. We have

$$\int_{-1}^{1} L_m(z)L_n(z)dz = 2/(2n+1) \text{ or } 0$$

according as m and n are equal or not; also the asymptotic formula

$$L_n(z) \sim (2\pi n)^{-1/2}(z^2-1)^{-1/4}[z+(z^2-1)^{1/2}]^{n+1/2}$$

is valid uniformly on any closed set having no point in common with E.

8.5.17. Let E and $f(z)$ satisfy the hypothesis of Ex. 8.5.15, and let $p_n(z)$ be the polynomial which interplates to $f(z)$ in the n+1 roots of the polynomial $Q_{n+1}(z)$, orthogonal on E with respect to a suitable weight function. Then we have

$$\int_{-1}^{+1} |f(z) - p_n(z)|^2 dz \leq \frac{M}{n^{2k+2\alpha}\ \rho^{2n}},$$

where M is independent of n and z. [Walsh and Sewell, 1940]

SUGGESTION. It is known that $|Q_n(z)| > M_1 \rho^n > 0$, z on C_ρ.

8.5.18. Let E satisfy the hypothesis of Ex. 8.5.15, and let f(z) be of class F_q on C_ρ. Let $p_n(z)$ be the polynomial which interpolates to f(z) in the n+1 roots of $T_{n+1}(z)$. Then we have

$$|f(z) - p_n(z)| \leq \frac{M}{n^{q-1/2}\ \rho^n}, \quad -1 \leq z \leq 1.$$

8.5.19. Corresponding to Exs. 8.5.15, 8.5.16, 8.5.18 we have results on approximation in the sense of least weighted p-th powers.

8.5.20. Let f(z) be analytic in $|z| < 1$, continuous on $|z| \leq 1$, and let $|f[e^{i\theta}] - f[(1-\delta)e^{i\theta}]| \leq \omega(\delta)$, where $\omega(\delta)$ is independent of θ. Then $p_n(z)$ exists such that $|f(z)-p_n(z)| \leq M\omega(2 \log n/n)$, $|z| \leq 1$. [Walsh]

8.5.21. Theorem 6.2.5 where C is the segment $-1 \leq z \leq 1$, and the points of interpolation are the roots of $T_{n+1}(z) = 2^{n-1}\cos(n \cos^{-1} z)$. [Walsh and Sewell, 1941]

8.5.22. In Theorem 6.2.7 let C be the segment $-1 \leq z \leq 1$. Then we have

$$|f'(z) - p_n'(z)| \leq M \epsilon_n n^2/\rho^n, \quad -1 \leq z \leq 1.$$

[Walsh and Sewell, 1941]

8.5.23. Let C be a rectifiable Jordan curve of Type t, $1 \leq t \leq 2$, containing the origin in its interior. Then $\int_C |p_n(z, 1/z)|^p |dz| \leq L^p$, $p > 0$, implies $|p_n(z,1/z)| \leq KLn^{t/p}$, z on C.

8.5.24. The conclusions of Exs. 8.5.1, 8.5.3, 8.5.5, and 8.5.7 hold if the circle $|z| = 1$ is replaced by an arbitrary analytic Jordan curve C containing the origin in its interior. [Compare Sewell, 1938b]

8.5.25. Let C be a Jordan curve of Type t containing the origin in its interior and let $f(z)$ be defined on C. Let $p_n(z,1/z)$ exist such that

$$|f(z) - p_n(z,1/z)| \leq \frac{M}{n^{(k+\alpha)t}}, \quad z \text{ on } C.$$

Then $f^{(k)}(z)$ exists on C and satisfies on C condition (8.5.3). [Compare Sewell, 1938b]

8.5.26. Let C be a rectifiable Jordan curve of Type t containing the origin in its interior and let $f(z)$ be defined on C. Let $p_n(z,1/z)$ exist such that

$$\int_C \Delta(z)|f(z) - p_n(z,1/z)|^p |dz| \leq \frac{M}{n^{[(k+\alpha)p+1]t}}, p > 0.$$

Then $f(z) = f_1(z)$ almost everywhere on C, where $f_1(z) = \lim_{n\to\infty} p_n(z,1/z)$; furthermore $f_1^{(k)}(z)$ exists on C and satisfies condition (8.5.3).

8.5.27. Let E be the segment $-1 \leq z \leq 1$; let $f(z) = (\rho - z)^h$, $\rho > 1$. Then $p_n(z)$ exists such that

$$|f(z) - p_n(z)| \leq \frac{M}{n^{h+1}\rho^n}, \quad -1 \leq z \leq 1.$$

[Bernstein, 1926, pp. 120-123]

8.5.28. Let E be the segment $-1 \leq z \leq 1$; let $f(z) = \log(\rho - z)$, $\rho > 1$. Then $p_n(z)$ exists such that

$$|f(z) - p_n(z)| \leq \frac{M}{n\rho^n}, \quad -1 \leq z \leq 1.$$

[Bernstein, 1926, loc. cit.]

8.5.29. Let $f(z)$, $z = x + iy$, be periodic of period 2π and analytic in the band $|y| \leq \log \rho$ except for a finite number of poles on the lines $y = \pm \log \rho$, each of order h at most. Then $t_n(z)$ exists such that for all real z

$$|f(z) - t_n(z)| \leq \frac{Mn^{h-1}}{n};$$

indeed $t_n(z)$ is the sum of the first $2n + 1$ terms of the Fourier development of $f(z)$. [de la Vallée Poussin, 1919, Chap. VIII]

8.5.30. Investigate necessary and sufficient conditions for functions of various classes in terms of the Laurent, Fourier, and the Legendre coefficients. [Walsh]

§8.6. DISCUSSION. We might have appropriately included in this chapter a section analogous to §6.5. In this connection the reader is referred to Exs. 8.5.27, 8.5.28, 8.5.29, also to Bernstein [1926] and de la Vallée Poussin [1919]. Such problems are classical but many questions are still unanswered.

It is to be observed that many of the results for the circle and the annulus are generalized in the exercises to a curve and the corresponding ring-shaped region. The reader should state and prove the theorems of Problem β corresponding to Exs. 8.5.24, 8.5.25, and 8.5.26 along with the extension of the definition of $L(k, \alpha)$ and $L'(k, \alpha)$ for ring-shaped regions. We assume here that the curve contains the origin and approximate by polynomials in z and $1/z$. For a curve arbitrarily situated in the plane it is natural to translate the origin into the interior of the curve, or to approximate by polynomials in $z-a$ and $1/(z-a)$ where a is a point interior to the curve. We are thus led to approximation by rational functions, which is a natural generalization of approximation by polynomials in z and $1/z$, or in $z-0$ and $1/(z-0)$. This affords a relatively new [compare Sewell, 1939c] and interesting study for which many of the preliminaries have been completed.

There are many other open problems which the reader will notice as he peruses the results of §§8.1, 8.2, 8.3 and the exercises relating thereto.

We turn now to §8.4. It is surprising indeed that purely elementary methods suffice to prove a result such as Theorem 8.4.1. Also the form of $p_n(z)$ in this theorem is interesting; it is closely related to the Taylor sum and hence the Fourier sum. In fact this yields an interesting result on trigonometric approximation, and one which can be proved by elementary methods, which is rare indeed in this theory. An investigation of the relation between $p_n(z)$ and $s_n(z)$ might well lead to an elementary proof of the classical theorem on the degree of approximation of the Fourier development.

With this we close our discourse. The results on degree of approximation on an analytic Jordan curve are fairly complete. It is the author's hope that the methods and results will serve others to advance this theory beyond the analytic frontier to much more hazardous regions, hitherto unruffled by the mind or at least the pen of mathematicians.

BIBLIOGRAPHY

AHLFORS, L. V.

1930. Untersuchungen zur Theorie der konformen Abbildung und der ganzen Funktionen. Acta Societatis Scientiarum Fennicae, Nova Series A, vol. 1, pp. 5-40.

BERGMAN, St.

1922. Über die Entwicklung der harmonischen Funktionen der Ebene und des Raumes nach Orthogonalfunktionen. Mathematische Annalen, vol. 86, pp. 238-271.

BERNSTEIN, S.

1912. Sur l'ordre de la meilleure approximation des fonctions continue par des polynomes de degré donné. Mémoires de l'Académie Royale de Belgique, (2), vol. 4, pp. 1-104.

1926. Leçons sur les propriétés extrémales et la meilleure approximation des fonctions analytiques d'une variable réelle. Paris.

BOCHNER, S.

1922. Über orthogonale Systeme analytischer Funktionen. Mathematische Zeitschrift, vol. 14, pp. 180-207.

CARATHÉODORY, C.

1932. Conformal Representation. Cambridge Tracts in Mathematics and Mathematical Physics, No. 28.

CARLEMAN, T.

1922. Über die Approximation der analytischer

Funktionen durch lineare Aggregate von vorgegebenen Potenzen. Arkiv för Matematik, Astronomi och Fysik, vol. 17, No. 9.

CURTISS, J. H.

1935. Interpolation in regularly distributed points. Transactions of the American Mathematical Society, vol. 38, pp. 458-473.

1936. A note on degree of polynomial approximation. Bulletin of the American Mathematical Society, vol. 42, pp. 873-878.

1941. On the Jacobi series. Transactions of the American Mathematical Society, vol. 49.

DARBOUX, G.

1876. Sur les développements en série des fonctions d'une seule variable. Journal de Mathématiques Pures et Appliquées, vol. II, 3^{e}Série, pp. 291-312.

DIENES, P.

1931. The Taylor Series. Oxford.

FABER, G.

1903. Über polynomische Entwickelungen. Mathematische Annalen, vol. 57, pp. 389-408.

1920. Über Tchebyscheffsche Polynome. Journal für die reine und angewandte Mathematik, vol. 150, pp. 79-106.

1922. Über nach Polynomen fortschreitenden Reihen. Sitzungsberichte der mathematisch-physikalischen Klasse der Bayerischen Akademie der Wissenschaften zu München, pp. 157-178.

FEJÉR, L.

1918. Interpolation und Konforme Abbildung. Nachrichten von der Gesellschaft der Wissenschaften zu Göttingen, Mathematisch-Physikalische Klasse,

pp. 73-86.

FEKETE, M.

1926. Über Interpolation. Zeitschrift für angewandte Mathematik und Mechanik, vol. 6, pp. 410-413.

GERONIMUS, J.

1931. On Orthogonal polynomials. Transactions of the American Mathematical Society, vol. 33, pp. 322-328.

1940. On the orthogonality of a system of polynomials on several curves (in Russian with English resumé). Communications de l'Institut des Sciences Mathématiques et Mécaniques de l'Université de Kharkoff et de la Société Mathématique de Kharkoff, vol. 16, ser. 4, pp. 12-32.

HADAMARD, J.

1901. La Série de Taylor. Scientia, Paris.

HARDY, G. H., AND LITTLEWOOD, J. E.

1928. Some properties of fractional integrals, I. Mathematische Zeitschrift, vol. 27, pp. 565-606.

1932. Some properties of fractional integrals, II. Ibid., vol. 34, pp. 403-439.

1935. An inequality. Ibid., vol. 40, pp. 1-40.

HEUSER, PAUL

1939. Zur Approximation analytischer Funktionen durch Polynome. Ibid., vol. 45, pp. 146-154.

HILLE, EINAR. See SHOHAT, J., HILLE, E., AND WALSH, J.L.

JACKSON, DUNHAM

1911. Über die Genauigkeit der Annäherung stetiger Funktionen durch ganze rationale Funktionen. Inaugural Dissertation, Göttingen.

1912. Approximation by trigonometric sums and polynomials. Transactions of the American Mathematical Society, vol. 13, pp. 491-515.

1914. On the degree of convergence of Sturm-Liouville Series. Ibid., vol. 15, pp. 439-466.

1930. The Theory of Approximation. American Mathematical Society Colloquium Publications, vol. XI.

1930a. On certain problems of approximation in the complex domain. Bulletin of the American Mathematical Society, vol. 36, pp. 851-857.

1931. On the application of Markoff's theorem to problems of approximation in the complex domain. Ibid., vol. 37, pp. 883-890.

1933. Certain problems of closest approximation. Ibid., vol. 39, pp. 889-906.

1933a. Orthogonal trigonometric sums. Annals of Mathematics, vol. 34, pp. 799-814.

JULIA, G.

1926. Sur les polynomes de Tchebycheff. Comptes Rendus hebdomadaires des Séances de l'Académie des Sciences (Paris), vol. 182, pp. 1201-1202.

KELDYSCH, M., AND LAVRENTIEFF, M.

1937. Sur la représentation conforme des domaines limités par des courbes rectifiables. Annales de l'École Normale, vol. 54, pp. 1-38.

KELLOGG, O. D.

1912. Harmonic functions and Green's integral. Transactions of the American Mathématical Society, vol. 13, pp. 109-132.

LAVRENTIEFF, M. See KELDYSCH, M., AND LAVRENTIEFF, M.

LINDELÖF, E.

1915. Sur un principe générale de l'analyse. Acta Societatis Scientiarum Fennicae, vol. 46, No.4, pp. 1-35.

See Also PHRAGMÉN, E., AND LINDELÖF, E.

LITTLEWOOD, J. E. See HARDY, G. H., AND LITTLEWOOD, J.E.

LIOUVILLE, J.

1832. Sur quelques questions de géométrie it de mécanique, et sur un nouveau genre de calcul pour résoudre ces questions. Journal de l'École Polytechnique, 1e Série, vol. 13, No. 21, pp. 1-69.

MANDELBROJT, S.

1927. Modern researches on the singularities of functions defined by Taylor's Series. Rice Institute Pamphlet, vol. 14, pp. 223-352.

1932. La singularité des fonctions analytiques représentées par une série de Taylor. Memorial des Sciences Mathématiques, vol. 54.

MARKOFF, A.

1889. Sur une question posée par Mendelieff. Bulletin of the Academy of Sciences of Saint Petersburg, vol. 62, pp. 1-24.

MERRIMAN, G. M.

1938. Concerning sets of polynomials orthogonal simultaneously on several circles. Bulletin of the American Mathematical Society, vol. 44, pp. 57-69.

See also WALSH, J. L., AND MERRIMAN, G. M.

MONTEL, PAUL

1910. Leçons sur les Séries de Polynomes à une Variable Complexe. Paris.

1919. Sur les polynomes d'approximation. Bulletin de la Société Mathématique de France, vol. 46, pp. 151-196.

OSGOOD, W. F., AND TAYLOR, E. H.

1913. Conformal transformations on the boundaries of their regions of definition. Transactions of the American Mathematical Society, vol. 14, pp. 277-298.

PHRAGMÉN, E., AND LINDELÖF, E.
1908. Sur une extension d'une principe classique de l'analyse. Acta Mathematica, vol. 31, pp. 381-406.

PÓLYA, G., AND SZEGÖ, G.
1925. Aufgaben und Lehrsätze aus der Analysis. Berlin.

PRIVALOFF, J.
1916. Sur les fonctions conjugées. Bulletin de la Société Mathématique de France, vol. 44, pp. 100-103.

RIEMANN, B.
1892. Gesammelte Mathematische Werke und Wissenschaftlicher Nachlass. Leipzig.

RIESZ, F.
1923. Über die Randwerte einer analytischen Funktion. Mathematische Zeitschrift, vol. 18, pp. 87-95.

RIESZ, M.
1914. Eine trigonometrische Interpolationsformel und einige Ungleichungen für Polynome. Jahresbericht der Deutscher Mathematiker Vereinigung, vol. 23, pp. 354-368.
1916. Über einen Satz des Herrn Sérge Bernstein. Acta Mathematica, vol. 40, pp. 337-347.

SAKS, S.
1937. Theory of the Integral. Monografje Matemalyczne, Warsaw-Lwów.

SEIDEL, W.
1931. Über die Ränderzuordnung bei konformen Abbildungen. Mathematische Annalen, vol. 104,

pp. 182-243.

SEWELL, W. E.

1935. Degree of approximation by polynomials to continuous functions. Bulletin of the American Mathematical Society, vol. 41, pp. 111-117.

1935a. Generalized derivatives and approximation. Proceedings of the National Academy of Sciences, vol. 21, pp. 255-258.

1936. On the modulus of the derivative of a polynomial. Bulletin of the American Mathematical Society, vol. 42, pp. 699-701.

1937. Generalized derivatives and approximation by polynomials. Transactions of the American Mathematical Society, vol. 41, pp. 84-123.

1937a. Note on the relation between integral and Tchebycheff approximation in the complex domain. Bulletin of the American Mathematical Society, vol. 43, pp. 425-431.

1937b. Degree of approximation by polynomials -- Problem α Proceedings of the National Academy of Sciences, vol. 23, pp. 491-493.

1937c. Note on the Faber coefficients of a continuous function. Revista de Ciencias, vol. 39, pp. 79-82.

1937d. On the polynomial derivative constant for an ellipse. American Mathematical Monthly, vol.44, pp. 577-578.

1938. The derivative of a polynomial on various arcs of the complex domain. National Mathematics Magazine, vol. 12, pp. 167-170.

1938a. Note on the relation between Lipschitz conditions and degree of polynomial approximation. Tôhoku Mathematical Journal, vol. 44, pp. 347-350.

1938b. Degree of approximation by polynomials in z and 1/z. Duke Mathematical Journal, vol. 4,

pp. 393-400.

1939. Jackson summation of the Faber development. Bulletin of the American Mathematical Society, vol. 45, pp. 187-189.

1939a. Integral approximation and continuity. Tôhoku Mathematical Journal, vol. 46, pp. 75-78

1939b. The derivative of a polynomial on further arcs of the complex domain. American Mathematical Monthly, vol. 46, pp. 644-645.

1939c. Continuity and degree of approximation by rational functions. Revista de Ciencias, vol. 41, pp. 435-451.

See also WALSH, J. L., AND SEWELL, W. E.

SHOHAT, J.

1933. On interpolation. Annals of Mathematics, 2nd Series, vol. 34, pp. 130-146.

1934. Théorie générale des polynomes de Tchebichef. Memorial des Sciences Mathématiques, vol. 66.

SHOHAT, J., HILLE, E., AND WALSH, J. L.

1940. Bibliography on Orthogonal Polynomials.

SMIRNOFF, V. J.

1928. Sur la théorie des polynomes orthogonaux à une variable complexe. Journal de la Société Physico-mathematique de Leningrad, vol. 2, pp. 155-178.

1932. Sur les formules de Cauchy et de Green et quelques problèmes qui s'y rattachent. Bulletin de l'Académie des Sciences de l'U.R.S.S., vol. 7, pp. 337-371.

SZASZ, O.

1922. Über den Konvergenzexponenten der Fourierschen Reihen gewisser Funktionenklassen. Sitzungsberichte der mathematisch-physikalischen Klasse der Bayerischen Akademie der Wissenschaften zu

München, pp. 135-150.

SZEGÖ, G.

1925. Über einen Satz von A. Markoff. Mathematische Zeitschrift, vol. 23, pp. 45-61.

1928. Zur Theorie der schlichten Abbildungen. Mathematische Annalen, vol. 100, pp. 188-211.

1935. A problem concerning orthogonal polynomials. Transactions of the American Mathematical Society, vol. 37, pp. 196-206.

1939. Orthogonal Polynomials. American Mathematical Society Colloquium Publications, vol. XXIII.

1939a. Concerning sets of polynomials orthogonal simultaneously on several circles. Bulletin of the American Mathematical Society, vol. 45, pp. 129-132.

See also PÓLYA, G., AND SZEGÖ, G.

TAYLOR, E. H. See OSGOOD, W. F., AND Taylor, E. H.

TONELLI, L.

1908. I polinomi d'approssimazione di Tchebychev. Annali di Matematica, (3), vol. 15, pp. 47-119.

VISSER, C.

1932. Über beschränkte analytische Funktionen und die Randverhältnisse bei konformen Abbildungen. Mathematische Annalen, vol. 105, pp. 28-39.

VALLÉE POUSSIN, C. J. de la

1914. Cours d'Analyse. Paris.

1919. Leçons sur l'Approximation des Fonctions d'une Variable Réelle. Paris.

WALSH, J. L.

1934. Note on the orthogonality of Tchebycheff polynomials on confocal ellipses. Bulletin of the American Mathematical Society, vol. 40, pp. 84-88.

1935. Interpolation and Approximation by Rational Functions in the Complex Domain. American Mathematical Society Colloquium Publications, vol. XX.

1935a. Approximation by Polynomials in the Complex Domain. Mémorial des Sciences Mathématiques, vol. 73.

See also SHOHAT, J., HILLE, E., AND WALSH, J. L.

WALSH, J. L., AND MERRIMAN, G. M.

1937. Note on the simultaneous orthogonality of harmonic polynomials on several curves. Duke Mathematical Journal, vol. 3, pp. 279-288.

WALSH, J. L., AND SEWELL, W. E.

1937. Note on the degree of approximation to an integral by Riemann sums. American Mathematical Monthly, vol. 44, pp. 155-160.

1937a. Note on the relation between continuity and degree of polynomial approximation in the complex domain. Bulletin of the American Mathematical Society, vol. 43, pp. 557-563.

1938. Note on degree of trigonometric and polynomial approximation to an analytic function. Ibid., vol. 44, pp. 865-873.

1940. Sufficient conditions for various degrees of approximation by polynomials. Duke Mathematical Journal, vol. 6, pp. 658-705.

1940a. Note on degree of trigonometric and polynomial approximation to an analytic function, in the sense of least p-th powers. Bulletin of the American Mathematical Society, vol. 46, pp. 312-319.

1940b. On the degree of convergence of harmonic polynomials to harmonic functions. Bulletin of the American Mathematical Society, vol. 46, abstract No. 341, pp. 434.

1941. Degree of polynomial approximation to analytic functions -- Problem β. Transactions of the American Mathematical Society, vol. 49.

WARSCHAWSKI, S.

1932. Über einen Satz von O. D. Kellogg. Nachrichten von der Gesellschaft der Wissenschaften zu Göttingen, Mathematisch-Physikalische Klasse, pp. 73-86.

1932a. Über die Randverhalten der Ableitung der Abbildungsfunktion bei konformer Abbildung. Mathematische Zeitschrift, vol. 35, pp. 321-456.

1934. Bemerkung zu meiner Arbeit: Über die Randverhalten der Ableitung der Abbildungsfunktion bei konformer Abbildung. Ibid., vol. 38, pp. 669-683.

ZYGMUND, ANTONI

1935. Trigonometrical Series. Warsaw-Lwow.

Added in proof:

CURTISS, J. H.

1941a. Degree of polynomial approximation on a lemniscate. Bulletin of the American Mathematical Society, vol. 47, p. 34, abstract 17.

Ingram Content Group UK Ltd.
Milton Keynes UK
UKHW011242290523
422449UK00001B/88

9 780691 095721